# ELEMENTARY STATISTICAL CONCEPTS

**ROLENE B. CAIN**
Associate Professor of Mathematics,
Morehead State University

1972

**W. B. SAUNDERS COMPANY**
PHILADELPHIA · LONDON · TORONTO

W. B. Saunders Company: West Washington Square
Philadelphia, Pa. 19105

12 Dyott Street
London, WC1A 1DB

833 Oxford Street
Toronto 18, Ontario

Elementary Statistical Concepts　　　　　　　　　　　　　　　　　　　　　　　　SBN-0-7216-2238-0

© 1972 by W. B. Saunders Company. Copyright under the International Copyright Union. All rights reserved. This book is protected by copyright. No part of it may be reproduced, stored in a retrieval system, or transmitted in any form or by any means, electronic, mechanical, photocopying, recording, or otherwise, without written permission from the publisher. Made in the United States of America. Press of W. B. Saunders Company. Library of Congress catalog card number 78-183447.

Print No.:　9　8　7　6　5　4　3　2　1

# PREFACE TO THE INSTRUCTOR

This text is intended for junior or senior level high school classes or for a three semester hour college freshman course in introductory statistics for students who have minimal background in mathematics. If a shorter course is desired, any or all of the following chapters may be deleted without loss of continuity: 6, 10, 11, 12, 16, 18, 19, or 20.

The text is designed so that one chapter per day may be presented, giving the student a sense of accomplishment. It is suggested that only one-half of the problems in each chapter be assigned. With this view in mind, the problems have been designed so that either the odd-numbered or the even-numbered problems form a complete set to illustrate the basic ideas of each chapter. The only exceptions are Chapters 4, 21, and 28. In these chapters only, all problems should be done by all students.

Since a minimum mathematical background of the reader is assumed, care is taken to give detailed verbal explanations of all concepts, even the most elementary ones. The instructor may find it necessary to present some review of arithmetic concepts, such as multiplication and division of fractions and addition of signed numbers, as questions arise.

Many laboratory materials may be purchased directly from supply houses, such as LaPine Scientific Company, 375 Chestnut St., Nor-

## PREFACE TO THE INSTRUCTOR

wood, N.J. 07648; Howell Enterprises, Ltd., P.O. Box 176, Littlerock, California 93543; Edmund Scientific Co., 402 Edscorp Bldg., Barrington, N.J. 08007; and the Lansford Publishing Co., 2516 Lansford Avenue, San Jose, California 95125. In addition, films from the Continental Classroom presentation of "Probability and Statistics" are available from the AudioVisual Center, Indiana University, Bloomington, Indiana 47405.

Purchased materials are not required, however. In all cases, the instructor can construct the required materials from everyday items. For example, a "sampling urn" with "balls" might be a bucket with ping-pong balls or marbles; or it could be a jar with beads, or—if the instructor does not think they would disappear too quickly—with jellybeans or candy pieces. The paper disks mentioned in Chapter 21 can be made with a large paper punch and relatively heavy paper such as paper towels. Spinners are available from many types of children's games. For the Poisson sampling experiment, one could use ping-pong balls with numbers painted on them. They may be sampled randomly from a bucket or tub. Alternatively, one could use marbles or beads and make a correspondence between color and number (red=1, blue=2, and so forth). The following two possible distributions of numbers will be Poisson with a mean of "2".

| Number | Frequency | |
|---|---|---|
| | EITHER | OR |
| 0 | 27 | 14 |
| 1 | 54 | 27 |
| 2 | 54 | 27 |
| 3 | 36 | 18 |
| 4 | 18 | 9 |
| 5 | 8 | 4 |
| 6 | 2 | 1 |
| 7 | 1 | |

Many times students will be intrigued by certain experimental outcomes, and extra credit projects of the learn-by-doing type will result. The basic purpose of the text is to inspire students with a limited mathematical background to learn the vocabulary of statistics, to investigate sampling procedures, and to be able to apply basic decision-making precedures in their everyday lives.

# PREFACE TO THE STUDENT

Experience is the best teacher. Statistics is a tool to be used in our everyday life. This text has been developed around these two premises. It is designed to help you think logically and rationally and to build a framework within which you may make decisions based upon information you have gathered. Statistics is, essentially, a kind of "applied horsesense."

There are many reasons for learning more about the field of statistics. Perhaps, because of your aptitude and interest in mathematics, you have had the opportunity to elect a statistics course before graduating from high school. Or you may be one of the many students with minimal background in high school mathematics who must take a course in basic statistics as a general education requirement, or else as a degree requirement in such fields of application as psychology, sociology, biology, business, or economics. Often the unfamiliar concepts and vocabulary of statistics, coupled with necessary algebraic manipulations, pose a distinct problem.

This text is designed to be a very elementary introduction to the basic statistical ideas which will be valuable to the high school student, to serve as a freshman level general education course in applied mathematics, or to prepare the college student for a higher level statistics course. You will be introduced to the vocabulary of statistics,

## PREFACE TO THE STUDENT

to simple statistical principles and experimental methods, and to the basic postulates of probability through the various experiments that you will be asked to perform. Many of these experiments are similar to those which resulted in the development of statistics as a branch of applied mathematics.

Read each lesson carefully, do the exericses and experiments following the lesson to be sure you understand the concepts, and then, if you cannot summarize them precisely in your own words, memorize the definitions for the terms listed. As you study these ideas, notice how statistics are being used in magazines, on radio, on TV, and in the newspapers — statistics are being used to describe situations, to entice you to buy products, to capture your vote in elections, and sometimes to mislead you and influence your thinking. Statistics are powerful tools — make them work for you!

# CONTENTS

*chapter 1*
INTRODUCTION .................................................................... 1

*chapter 2*
DISTRIBUTIONS .................................................................. 13

*chapter 3*
AVERAGES ........................................................................ 19

*chapter 4*
SAMPLING FROM A POPULATION ..................................... 31

*chapter 5*
THE CONCEPT OF RANDOMNESS ..................................... 45

*chapter 6*
A ONE-SAMPLE TEST OF RANDOMNESS............................ 50

*chapter 7*
THE UNIFORM DISTRIBUTION .......................................... 55

## CONTENTS

*chapter 8*
**THE BERNOULLI DISTRIBUTION** ..................................................... 65

*chapter 9*
**THE BINOMIAL DISTRIBUTION** ........................................................ 72

*chapter 10*
**THE BINOMIAL TEST** ..................................................................... 85

*chapter 11*
**THE SIGN TEST** ............................................................................ 92

*chapter 12*
**SIGNED RANKS TEST** ................................................................... 101

*chapter 13*
**PERMUTATIONS AND COMBINATIONS** ............................................. 107

*chapter 14*
**APPLICATIONS OF COUNTING RULES** ............................................. 120

*chapter 15*
**INDEPENDENT EXPERIMENTS** ........................................................ 123

*chapter 16*
**THE MANN-WHITNEY $U$ TEST** ........................................................ 130

*chapter 17*
**MEASURES OF DISPERSION** .......................................................... 137

*chapter 18*
**A TEST FOR DISPERSION** .............................................................. 148

*chapter 19*
**MEASURES OF SKEWNESS AND KURTOSIS** ..................................... 157

*chapter 20*
**THE WALD-WOLFOWITZ RUNS TEST** ............................................... 170

*chapter 21*
**THE POISSON DISTRIBUTION AND ITS APPLICATIONS** ..................... 174

*chapter 22*
A TEST OF FIT—THE CHI-SQUARE DISTRIBUTION ............................ 184

*chapter 23*
THE MEDIAN TEST................................................................................ 201

*chapter 24*
ESTIMATION ........................................................................................... 205

*chapter 25*
DECISION MAKING................................................................................ 220

*chapter 26*
THE NORMAL DISTRIBUTION............................................................. 224

*chapter 27*
CASE STUDIES ...................................................................................... 237

*chapter 28*
A RESEARCH SURVEY ........................................................................ 242

APPENDICES .......................................................................................... 245

INDEX........................................................................................................ 265

chapter 1

# INTRODUCTION

We are bombarded by statistical results from all directions these days. There are baseball batting averages and "statistics" reporting hits, runs, and errors. TV newcasts flash to us the results of investigations into the relationship of cigarette smoking to heart disease and lung cancer. On radio and TV, weather forecasts specify the "probability of rainfall" and the "mean daily temperature." In newspaper articles we encounter analyses of the rise and fall of consumer indexes, gross national product figures, and cost of living indexes. We need to know enough about statistics to be able to evaluate critically the statements made in news articles, advertisements, or radio and TV commercials so that we are not led (or misled) blindly by them.

Where there are statistics there must be a statistician. Young people planning their futures should investigate this career. The academic world, industry, private consulting, and government work all demand capable, trained statisticians.

What does a statistician do? Some statisticians teach statistics in schools and colleges. Others are involved in market research, designing tests for existing products to learn which brands are best for certain purposes. Many analysts survey consumers to learn what new products their firms should produce for the market. Some statisticians are quality control specialists. Others work for the Federal Government in the Bureau of Census in Washington. Many act as consultants for firms which do not need a full-time statistician but which occasionally

## INTRODUCTION

have statistical problems to solve. Many statisticians serve in the field of economics, appraising the current business outlook; they advise their firms about such things as the expansion of facilities, the most efficient system of distribution of finished products, sales or purchases of capital equipment, scheduling of factory production to stabilize employment for seasonal products, and so forth. People interested in studying birth defects, epidemics, or the effects of new drugs on "incurable" diseases become biostatisticians.

Insurance companies determine policy rates and liability insurance risks by statistical analyses of accident, disease, and death records. Some statisticians work for state or local welfare agencies, conducting surveys of welfare recipients or perhaps doing cost analyses of social insurance. Space vehicles are returning a wealth of data which must be analyzed and interpreted by statisticians. In short, whenever there is the need to make a decision in the face of uncertainty—whether that decision be to carry an umbrella today or to proceed as planned on the next venture into space—statistics can assist the person in charge to make the best decision under a given set of circumstances.

In businesses, research institutions, and government offices, as well as in everyday life, we must often make decisions. We cannot be certain which is the best decision because we seldom have all the necessary facts at hand, or because the outcome of our decision depends to some extent on people and events beyond our control. For example, a businessman does not know what his competitors will do during the next few months or how future changes in government policy or in the general economic picture will affect his business; yet he must plan for the future and decide how and when to expand his business. Faced with a need to choose among alternatives, we can employ a knowledge of statistics to help us make the best decision under a given set of circumstances.

Before such decisions can be made wisely, we may need to perform experiments, conduct surveys, or in some other way gather information about things that could influence our decision. Before finalizing his expansion plans, the businessman mentioned in the previous paragraph would probably conduct surveys to find out how his customers felt about his products. Statistics could help him gather and analyze the necessary information.

Physical manipulation or special treatment of material, items, or subjects is called an **experiment.** Notations made concerning the behavior of the materials or subjects are called **observations.** If the observations themselves are not already numerical, then numbers are assigned to them to reflect the basic characteristics of the subjects' responses. This process is called **measurement.** Thus in a given experiment the observations taken are sometimes numbers, such as heights or weights of people; sometimes they are classifications, such as whether an experimental animal is male or female; sometimes they are ranks, such as whether an individual placed first, second, or third in a race. For analysis, however, some observation must be taken and some method is usually devised to assign a numerical measure to the observations so that the numbers meaningfully reflect the characteristics of the subjects' responses to treatment.

**Statistics** is the art of decision making under uncertainty. Practicing this art often involves collecting, manipulating, and summarizing large masses of data. From the time we arise in the morning until we go to bed at night, we are constantly faced with decisions. The simple decision of what to wear depends upon whether it is cold or hot and upon what we plan to do or whom we expect to meet. Sometimes we need to know how long it will be before we can have lunch in order to decide what we want for breakfast. The subject we major in depends upon what career interests us most; and then our major determines the courses we study in any particular school term. The list is endless. These decisions are usually made under uncertainty. Our various courses of action promote certain reactions or consequences which reveal whether our decision was good or poor.

**Descriptive statistics** enable the experimenter to classify a mass of experimental data into small, meaningful units or values. In newspapers we see pie diagrams showing the national budget, tables which show the results of opinion polls, maps of the United States showing areas that have varying degrees of shading to denote weather patterns or areas where certain commodities are produced, and so on. These are all examples of descriptive techniques. If the results of an experiment are used to infer properties about material that was not actually included in the experiment, then **inferential** or **inductive statistics** is involved.

EXAMPLE 1–1. A student uses the inductive method when he decides to take a course because his friends have taken the course and enjoyed it (or received a high grade or liked the teacher, etc.). Thus he has drawn a general conclusion (that he too will enjoy the course or like the teacher or receive a high grade) about similar material (himself) not actually included in the experiment (his friends, not he, took the course).

EXAMPLE 1–2. Suppose you go to a local hamburger stand for a snack after a movie and have to wait 25 minutes to get your order. If you tell your friends about having to wait, you are using a descriptive technique; but if you refuse to go to that stand again because you do not like to wait 25 minutes for your order—and thus assume you will have to wait next time, too—you are employing an inductive technique.

In summary, suppose a decision must be made and the person in charge feels that he does not have sufficient information or knowledge upon which to base his decision. He may design and conduct an experiment, or send out a questionnaire, or make a telephone survey to gather information. Merely describing the results may suffice to give

## INTRODUCTION

him the information he needs. On the other hand, he may wish to infer that other people not surveyed would feel the same way. Alternatively, he may wish both to describe the results and to infer that his results are characteristic of a general group of people. Thus, he may wish to use either descriptive or inferential statistical methods, or both. Similarly, in the situation of Example 1–2, you may tell your friends of the poor service at the hamburger stand, you may refuse to go there again, or you may do both. In any event, the methods of statistics facilitate the description and analysis of the data so that we may reach a decision utilizing all available information.

Often the methods used in analyzing or interpreting experimental data differ according to the **measurement scale** used. The **nominal scale** merely names categories, such as male-female, or success-failure, or Democrat-Republican-Independent. There may be any number of categories, but every item must be uniquely classified as belonging to one or another of them. If numbers are assigned, such as 1 = male and 2 = female, the numbers have no meaning as an actual numerical measure; in other words, it makes no sense to talk about 1.5 as being half-male and half-female. The **ordinal scale** is a ranking measure, such as private-corporal-sergeant, or first-second-third place in a race, or A-B-C-D-E in a grading scale. There still may be categories named, but with ordinal measure the categories now are ranked as being "better" or "worse," "higher" or "lower," etc. The **interval scale** is a true numerical measurement scale in which numerical differences have meaning, such as degrees for temperature, pounds for weight, feet for height, or counting measure for the number of redheads in a class. Sometimes it is difficult to classify certain observations as to measurement scale, and sometimes the classification will vary according to the context of the experiment. For example, the group assignment of children to reading groups would be nominal if the choice of any particular child for Group A, Group B, and so on were made arbitrarily. However, if the children were assigned according to their reading ability, it would be ordinal, since it would rank them. If, after the children are assigned, we wished to count the number of children in each group, our counts would involve the interval scale.

**Exercise 1–1:** Indicate by N, O, or I which measurement scale is involved in measuring each of the following:

_____ 1. Sex

_____ 2. University class (freshman, sophomore, etc.)

_____ 3. Days of the month of February

_____ 4. Degrees (centigrade) of temperature

_____ 5. Length of root in corn seed sprouts

# INTRODUCTION

_____ 6. Weight of tobacco leaves

_____ 7. Number of horses in a field

_____ 8. Grades on an English essay

_____ 9. Numbers on football jerseys

_____ 10. Serial numbers on cars

_____ 11. Heights of a group of university coeds

_____ 12. Hair colors of freshman students

_____ 13. Shoe sizes of members of a biology class

_____ 14. Political parties

_____ 15. Place awards in a cake-baking contest

**Exercise 1-2:** Indicate by N, O, or I which measurement scale is involved in measuring the following:

_____ 1. Religious affiliation (to what church does a person belong?)

_____ 2. Ages of children in a 4-H club

_____ 3. Number of people passing a certain street corner

_____ 4. Weights of rats

_____ 5. Makes of typewriters (IBM, Smith-Corona, Royal, etc.)

_____ 6. Positions on a baseball team

_____ 7. Amount of tar in different brands of cigarettes

_____ 8. Brands of cigarettes

_____ 9. Heights of file cabinets

_____ 10. Angle of inclination of the sun at certain times of day

_____ 11. "Win," "place," and "show" in a horse race

_____ 12. Energy of a body in motion

## INTRODUCTION

_____ 13. Position of Venus

_____ 14. Heights of athletes

_____ 15. Heat generated by a biological organism

Graphs can be used to illustrate the relationships among several measurements. Some types of graphs are particularly suited to certain measurement scales. If the objects compared are measured in the nominal scale, the bar chart is often used. This is a chart which uses vertical or horizontal bars to show relationships. For example, if a classroom contains 14 boys and only 8 girls, this fact can be illustrated by a bar chart as in Figure 1-1. To decide whether to use a horizontal or vertical bar chart, see whether the labels for the bars are long words (horizontal best) or short words (vertical best). In our example "girls" and "boys" are relatively short as labels. In such a case, we use whichever one looks best.

Either a polygon or a histogram might be used to describe the number of people in each age bracket among the 22 people in a classroom. The polygon is a line graph which merely connects the points indicating the total numbers, or **frequencies**, for each category. The histogram employs adjacent rectangles with heights corresponding to the category frequencies. Figure 1-2 shows a polygon and a histogram drawn from the same data. Both shapes show relationships among the categories or groups in terms of areas. The polygon is best when one wants to compare two groups of data on the same graph, such as the ages of people in a freshman English class compared to the ages of people in a senior history class. Both types of graphs are used with the interval measurement scale; and the continuity or numerical nature of the measurement scale is implied in the histogram by the fact that on the horizontal axis the individual class ranges are contiguous (the rectangles touch each other).

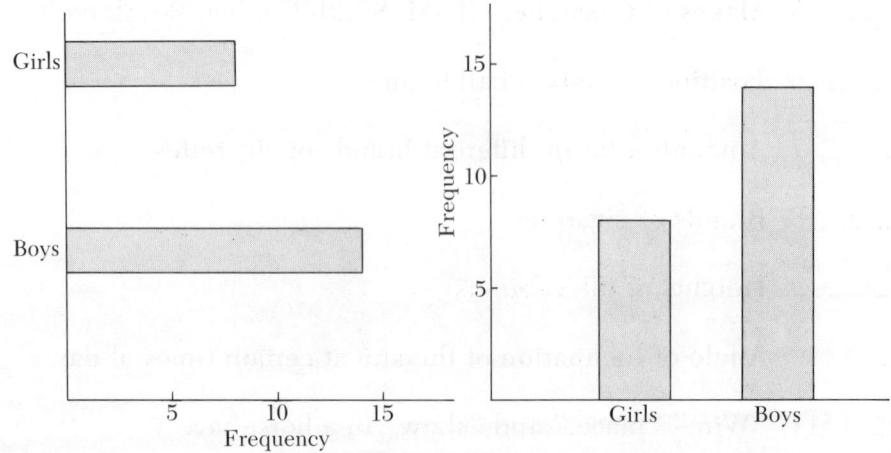

**Figure 1-1** Horizontal and vertical bar charts showing frequencies of boys and girls in a classroom.

# INTRODUCTION

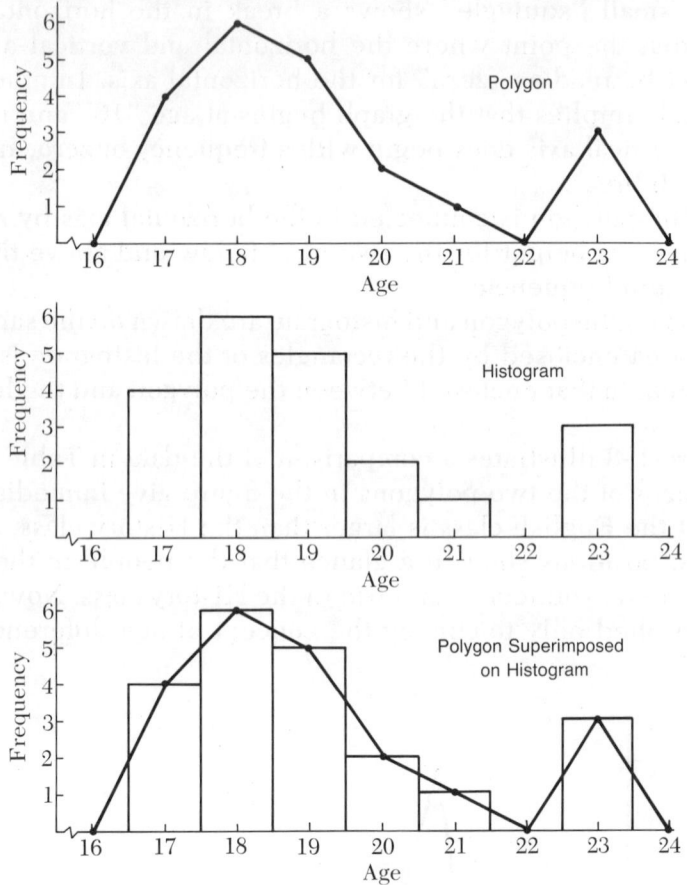

**Figure 1-2** Polygon and histogram compared.

The bar chart gives a similar type of area representation, but illustrates its underlying nominal or ordinal scale by having each bar separate and not contiguous. The use of a bar chart for nominal data and of a polygon, or histogram for interval data is not a hard and fast rule. Instead, it is merely a guide to the experimenter. One should always use that graphical method which best portrays what he wants to emphasize in his description of the experimental results.

From either the polygon or the histogram in Figure 1-2, we see that there were 4 seventeen year olds, 6 eighteen year olds, and so forth. Several properties are evident.

(a) The rectangles in the histogram are centered at the mark showing the age classification; each has a width of one unit, and its height is equal to the number of items in the category named by the value corresponding to the point at which the rectangle is centered. For the polygon the dot indicating the frequency of a specified age is placed directly over the mark showing the age classification.

(b) For a frequency of zero (such as for age 22), there is no rectangle (a rectangle of zero height), and there is also a dot at zero height in the polygon.

## INTRODUCTION

(c) A small "squiggle" shows a break in the horizontal axis to indicate that the point where the horizontal and vertical axes cross should not be read as "zero" for the horizontal axis. In other words, the squiggle implies that the graph begins at age "16" and not at age "0". The vertical axis does begin with a frequency of zero, and thus it has no such break.

(d) The polygon is connected to the horizontal axis by recording dots at the zero height for the two ages below and above those ages with non-zero frequencies.

(e) When the polygon and histogram are drawn on the same graph, the total area enclosed by the rectangles of the histogram is approximately equal to that enclosed between the polygon and the horizontal axis.

Figure 1-3 illustrates a comparison of the data in Table 1-1. The relative sizes of the two polygons in the figure give immediate visual proof that the English class is larger than the History class, and their respective positions show at a glance that the people in the English class tend to be younger than those in the History class. Now, suppose that we wished only to convey the concept of age difference and to

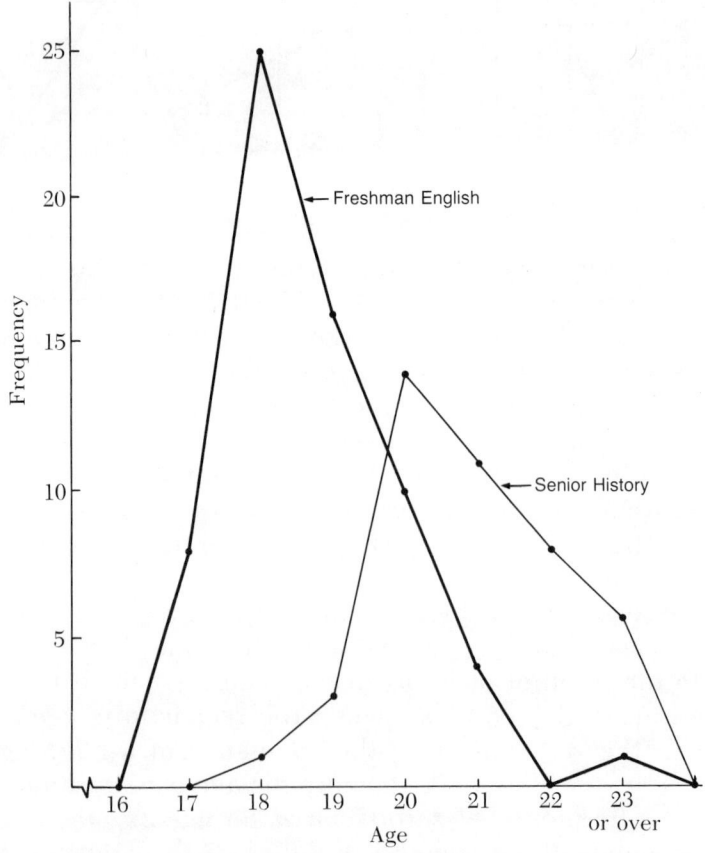

**Figure 1-3** Comparison of ages of people in two different classes by frequency polygons. The data are from Table 1-1.

**TABLE 1-1  AGES OF PEOPLE IN TWO DIFFERENT CLASSES**

|  | Freshman English | | Senior History | |
| --- | --- | --- | --- | --- |
| Ages | NUMBER | PROPORTION | NUMBER | PROPORTION |
| 17 | 8 | .125 | 0 | .000 |
| 18 | 25 | .391 | 1 | .024 |
| 19 | 16 | .250 | 3 | .072 |
| 20 | 10 | .156 | 14 | .333 |
| 21 | 4 | .062 | 11 | .262 |
| 22 | 0 | .000 | 8 | .190 |
| 23 or over | 1 | .016 | 5 | .119 |
|  | 64 | 1.000 | 42 | 1.000 |

ignore the difference in class sizes. We could then divide each observation by the total number of students in its class; this would give the proportion (or **relative frequency**) of people in each age group in the two classes. We could then graph the data as in Figure 1-4. This method would effectively illustrate the differences in ages between the classes and ignore the different class sizes.

A pie chart is the best method of graphical comparison. Nominal

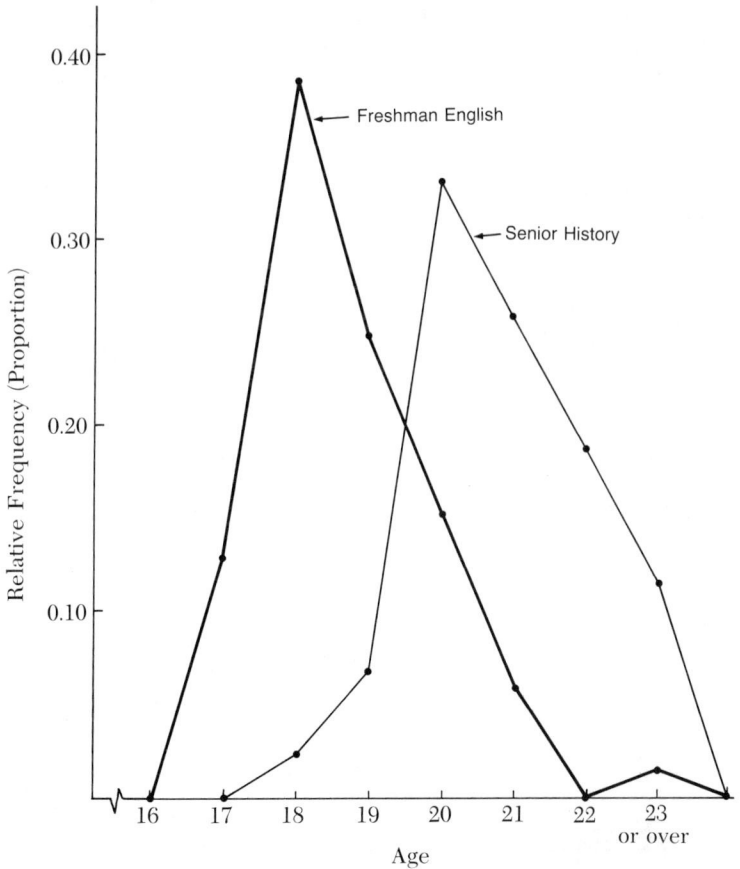

**Figure 1-4**  Comparison of ages of people in two different classes by relative frequency polygons.

## INTRODUCTION

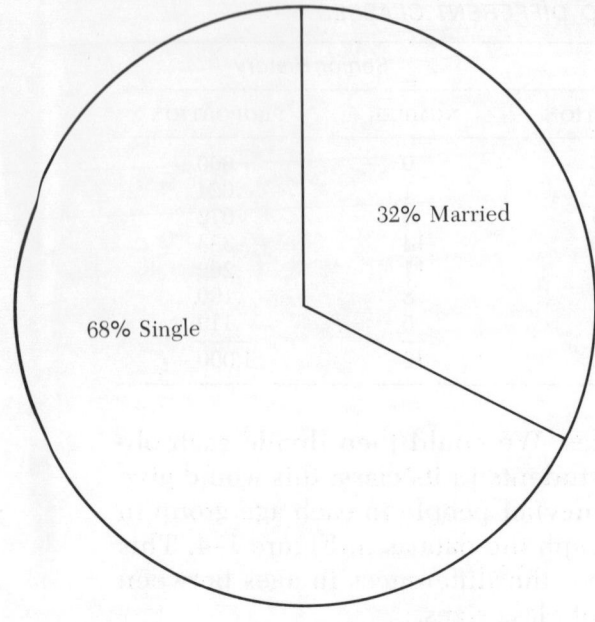

**Figure 1-5** Pie chart showing proportion of married students in a class of 22 students.

data are measured using proportions or percentages. Figure 1-5 gives an example of a pie chart comparing the number of married students with the number of unmarried students in a class of 22 people. A pie chart is very specialized, and should only be used for proportion data if there are no small proportions that would be difficult to see. Sometimes several small proportions can be grouped together to form a "miscellaneous" or "other" category.

***Exercise 1-3:*** Which type of chart or graph would you choose for the following situations? Why?

(a) To compare sales of Chevrolets in each month of 1960 with sales in each month of 1970. _____

_____

(b) To indicate how sales of Chevrolets in 1970 varied from month to month. _____

_____

(c) To compare total dollar sales in 1970 of Fords, Chevrolets, Buicks, and Lincolns. _____

_____

# INTRODUCTION

(d) To compare the proportion of total sales in 1970 accounted for by Fords, Chevrolets, Buicks, and Lincolns. _____

_____

**Exercise 1-4:** Which type of chart or graph would you choose for the following situations? Why?

(a) A cat, a dog, and a rat had all been trained to press a bar to obtain food after an electric light had been turned on in the room where they were caged. After it was evident that they had learned their bar-pressing task, an experiment was designed to compare the speed of reaction for each type of animal. Each was observed on ten occasions and its reaction time, or the time elapsed between the light and the pressing of the bar, was recorded. How may these three series of ten measurements be compared graphically? _____

_____

(b) One hundred rats are being trained to press a bar for food after an electric light comes on. At the end of each training period the number of rats who have learned the task are counted. Which graphical method would be best to show the relationships in these data? _____

_____

(c) Ten dogs, ten cats, ten rats, and ten monkeys have all been subjected to a test to measure their reaction times as outlined in (a) and an average reaction time for each has been obtained. How may these four averages be compared graphically? _____

_____

(d) In the experiment outlined by (b), the proportion of rats who had learned the task after the second, fourth, and sixth trials was calculated, and the remainder (who had not learned the task after six trials) were classified "other." How may these proportions be compared graphically? _____

_____

## INTRODUCTION

### TERMS TO REMEMBER

**Experiment:** Physical manipulation or special treatment of materials, items, or subjects.

**Observations:** Notations made concerning the resultant behavior of the materials, items, or subjects when they have been treated or manipulated in an experiment.

**Measurement:** The assigning of numbers to experimental observations in such a way that the numbers reflect some basic characteristic of the materials, items, or subjects after they have undergone some experimental treatment.

**Statistics:** The description and analysis of experimental data and the art of decision making under uncertainty.

**Descriptive Statistics:** Classification of a mass of data into tables, charts, or small units.*

**Inferential Statistics:** The process of using experimental data to draw general conclusions about the characteristics of similar material not actually included in the experiment.

**Measurement Scales:**
    (a) "Nominal" merely categorizes or names classifications.
    (b) "Ordinal" ranks the data, but differences in the numerical values assigned to the ranks are meaningless.
    (c) "Interval" is numerical measure for which equal interval differences have the same meaning.

**Frequency:** The number of objects in a category or class.

**Relative Frequency:** The proportion of items in one classification, found by dividing the number of items in one particular category by the total number observed in all categories.

---

*The "small units" are such things as means, medians, and standard deviations, which will be introduced in later chapters.

chapter 2

# DISTRIBUTIONS

Basic to any consideration of statistical methods is the concept of a **distribution,** which is a systematic arrangement of observations or objects into meaningful categories or classifications. In his everyday life a person tends to organize the things around him. The housewife distributes the furniture in a room so that the arrangement is visually pleasing and functional. When it is time to grant raises, the businessman ranks his employees as to their abilities and the relative values of their contributions. The child, given a handful of change, often stacks the pennies, dimes, nickels, and so on separately. Similarly, after obtaining a mass of experimental data, the researcher rearranges his observations according to some systematic scheme.

EXAMPLE 2-1. A child, before counting the money in his piggy bank, might arrange the coins and notice 2 quarters, 5 dimes, 16 nickels, and 21 pennies. He would probably stack these coins separately. The stacks could be described by a table listing the types of items with accompanying counts, or frequencies, of each type. This listing is called **frequency table,** and might look like Table 2-1.

**TABLE 2-1** FREQUENCY TABLE

| Type of Coin | Frequency | Proportion or Relative Frequency |
|---|---|---|
| quarter | 2 | 2/44 or .05 |
| dime | 5 | 5/44 or .11 |
| nickel | 16 | 16/44 or .36 |
| penny | 21 | 21/44 or .48 |
|  | 44 | 1 |

## DISTRIBUTIONS

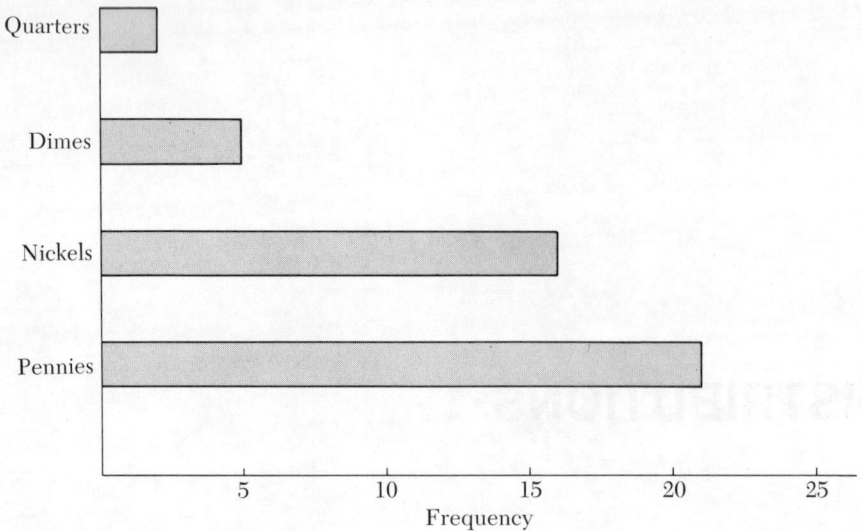

**Figure 2-1** Bar chart summarizing the frequencies of various coins in a handful of change.

When the coins are stacked or listed in a frequency table, it is much easier to count the total amount of money involved. Moreover, instead of 44 individual items, there are only 4 categories, each with a size measurement attached. A bar chart, such as the one in Figure 2–1, shows the relationship among the different categories of coins. From the chart we can see immediately that the penny occurs most often, that the quarters occur least often, that there are about three times as many nickels as dimes, and so forth. Table 2–1 and Figure 2–1 both show the *distribution* of coins in the piggy bank.

EXAMPLE 2-2. In a recent poll of 25 college students, each one was asked how much he spent for breakfast that morning. The following information resulted: .75, .53, 1.25, 1.05, .59, 0, .10, 1.50, .32, .45, .89, .72, .65, 1.10, 1.05, 1.14, 1.52, .90, 1.75, .05, 0, .85, .75, .20, 0. A table listing each number separately would not summarize the information at all. To summarize the data we should choose a few categories, each of which has the same range of values. The actual number of categories chosen is up to the researcher. In general, a few numbers (like 30 or fewer) would call for few categories (like 5 or 6), and many numbers (like several hundred) would be better summarized in a larger number of categories (like 15 to 18 or so). The ranges must be specified in such a way that there is no question about where any single observation should be placed. Since our smallest observation in this example is zero and the largest is 1.75, we could summarize the data using 5 categories as in Table 2–2. This is one distribution of the data; if we had chosen, for instance, to use 10 categories of .20 each, that would be another distribution, even though the data are the same.

To visualize what is meant by the term "distribution," do the

## DISTRIBUTIONS

**TABLE 2-2** AMOUNTS SPENT FOR BREAKFAST

| Range of Amount Spent | Frequency |
|---|---|
| .00 – .39 | 7 |
| .40 – .79 | 7 |
| .80 – 1.19 | 7 |
| 1.20 – 1.59 | 3 |
| 1.60 – 1.99 | 1 |
|  | 25 |

following exercises. They illustrate how items of data can be distributed into categories, counted and then summarized by a frequency table and graph. Use a group of about twenty people as experimental subjects.

**Exercise 2-1:** Record the distribution of eye colors in the following table:

| Color | Frequency |
|---|---|
| Brown |  |
| Blue |  |
| Hazel |  |
| Green |  |
| Gray |  |
| Other |  |

Next graph the data (using a bar chart, since these are nominal data).

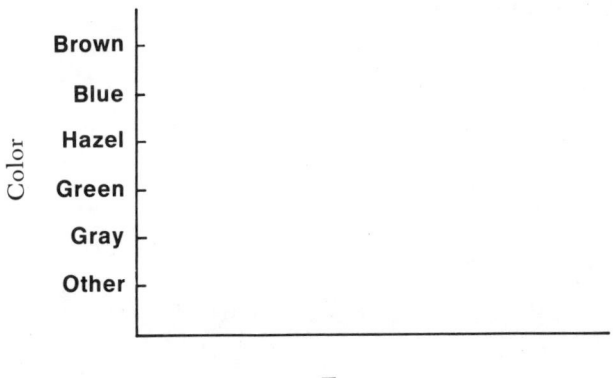

15

## DISTRIBUTIONS

**Exercise 2–2:** Obtain the distribution of footwear and record it in the following table:

| Type of Footwear | Frequency |
|---|---|
| oxfords | |
| loafers | |
| sandals | |
| boots | |
| high heels | |
| tennis shoes | |
| barefoot | |

Next graph the data.

**Exercise 2–3:** Obtain the distribution of shoe size (separately for each sex) and graph it.

**Exercise 2-4:** Determine and graph the distribution of money spent per week attending the movies. (Again, it may be instructive to separate the observations by sexes.)

**Exercise 2-5:** Determine and graph the distribution of hours spent per week watching television.

**Exercise 2-6:** Determine and graph the distribution of hours spent per week doing homework.

# DISTRIBUTIONS

### TERMS TO REMEMBER

**Distribution:** A systematic arrangement of observations into meaningful classifications or categories.

**Frequency Table:** A listing of categories or measurement ranges, with the frequency or total number of times a member of each category was observed.

chapter 3

# AVERAGES

Most people are familiar with the process of averaging several observations by adding them and dividing by their number. For example, if there are twenty people in one statistics class, thirty-five in a second class, and forty-one in a third, then the average number of people per class is $\frac{20 + 35 + 41}{3}$, or 32. This type of average is called the **mean**. It is often used with interval data that do not have a few extremely large or extremely small observations, called **outliers**.

EXAMPLE 3–1. Sometimes the mean does not adequately measure the central tendency in data with outliers. Consider this situation: the incomes of a plantation owner, his son, and his 98 laborers are, respectively, $1000, $750, and $75 per week. The mean income is $\frac{1000 + 750 + 98(75)}{100} = 91$. If the plantation owner announced $91 as the "average income" of those working on his plantation, each one of the 98 workers would be convinced that his 97 co-workers each received $16 per week more than he did, and the owner would be faced with a revolt. A much more meaningful measure of the location of the bulk of the observations is the value $75, as we shall see shortly.

The mean is not the only kind of "average." There are several

# AVERAGES

others which, together with the mean, are called **measures of location** or **measures of central tendency** of a set of data. Each one has certain particular uses. In any given experimental situation the experimenter decides which type of central measure is best for his purposes.

The **median,** another of the location measures, is used with ordinal data and with interval data that have outliers. To obtain the median, we first put the observations in ascending or descending order and then identify the middle number as the median. In Example 1, the median is $75. If only grouped data are available, we draw a histogram of the data and find the value that divides the area of the histogram exactly in half. Sometimes the median could be any number in an interval; in that case, the midpoint of the interval is taken as the median, as in Example 3-2.

EXAMPLE 3-2. To obtain the median of the experimental observations 1,7,2,14,3, and 10, we first list them in order: 1,2,3,7,10,14. Notice that any number between "3" and "7" divides them in half. Thus the median is the midpoint between "3" and "7" or $\frac{3+7}{2} = 5$.

EXAMPLE 3-3. Suppose the observations, listed in ascending order, were 1,2,2,2,4,6. Here the median is 2, since the middle two numbers are both "2".

EXAMPLE 3-4. If the observations are 47,53,62,71, and 39, then the median is 53.

EXAMPLE 3-5. Consider the data discussed in Example 2-2. Since we know what each of the 25 students spent for breakfast, we can compute the median of these observations by listing the amounts in increasing order and then selecting the 13th number in the list; the median is 0.75. But suppose instead that only the grouped data of Table 2-2 were available. We may summarize the data with a frequency table and histogram as shown in Table 3-1.

Then we must find the point that divides the area of the histogram in half, with "12.5" students to the left and "12.5" to the right of that point. This value would occur somewhere in the rectangle that is centered at .60. To find the exact point, notice that we have 7 of the students to the left of the median point grouped in the category of 0 to .39. Thus, in order to get "12.5" students to the left of the median, we need only 5.5 of the 7 units that lie between 0.40 and .79 (or, to be exact, between .40 and the lower value of the next category, namely .80). Thus, we begin with .40, which is the lower endpoint of the interval, and add to that $\frac{5.5}{7}$ times the length of the interval. In other words, the median equals $.40 + \frac{5.5}{7}(.80 - .40) = .714$. This is not the same as .75, the value we obtained from the individual observations. Whenever we calculate statistical measures from grouped data, we lose precision. The correct median is .75, and .714 is only an estimate of the median based on grouped data.

**TABLE 3-1  AMOUNTS SPENT FOR BREAKFAST**

| Range of Amount Spent | Frequency |
|---|---|
| .00 – .39 | 7 |
| .40 – .79 | 7 |
| .80 – 1.19 | 7 |
| 1.20 – 1.59 | 3 |
| 1.60 – 1.99 | 1 |
| | 25 |

Amounts Spent for Breakfast

In summary, the steps necessary to calculate the median from grouped data are:
1. Divide the total number of observations by 2; call the result $M$.
2. Find the class that contains the $M$th observation. Call it the median class.
3. Find the lower boundary of the median class. Call it $L_1$.
4. Find the lower boundary of the class immediately above the median class. Call it $L_2$.
5. Let $f$ be the frequency in the median class.
6. Subtract the total of the frequencies of all the classes below the median class from $M$ and call the result $j$.
7. The median is then $L_1 + \dfrac{j}{f}(L_2 - L_1)$.

A third type of average is the **mode**. It is the observation that occurs most often, and is quickly and easily found. The mode is the

# AVERAGES

only average that can be used when the observations have nominal measure.

EXAMPLE 3-6. The following sandwiches were ordered at the Dairy Queen between 6 and 7 P.M. one evening: hamburger, foot-long, hamburger, cheeseburger, hotdog, cheeseburger, hamburger, hotdog, hamburger. The typical or "average" sandwich was hamburger, since it occurred most often. Since the observations of "hamburger", "foot-long", and so on have nominal measure, the mode is the appropriate measure of central tendency.

EXAMPLE 3-7. In the sets of numbers given in Examples 3-2 and 3-4, there are no modes, since for each, no number occurs more often than any other. In Example 3-1 the mode is "$75", and in Example 3-3 the mode is "2". Sometimes there is not a unique mode: the set of observations 2,2,3,5,7,7,8,10,12,12, would be called "trimodal" since it has three modes, the numbers 2,7, and 12.

Finally, to locate the center of the range of the observations, we employ the **midrange.** It can be used only with interval data. It is equal to half the sum of the largest and smallest observations. The midrange is easy to calculate and especially useful whenever many repeated observations are difficult to take, such as with daily temperatures or with stock prices; however, it has the disadvantage of being affected by outliers.

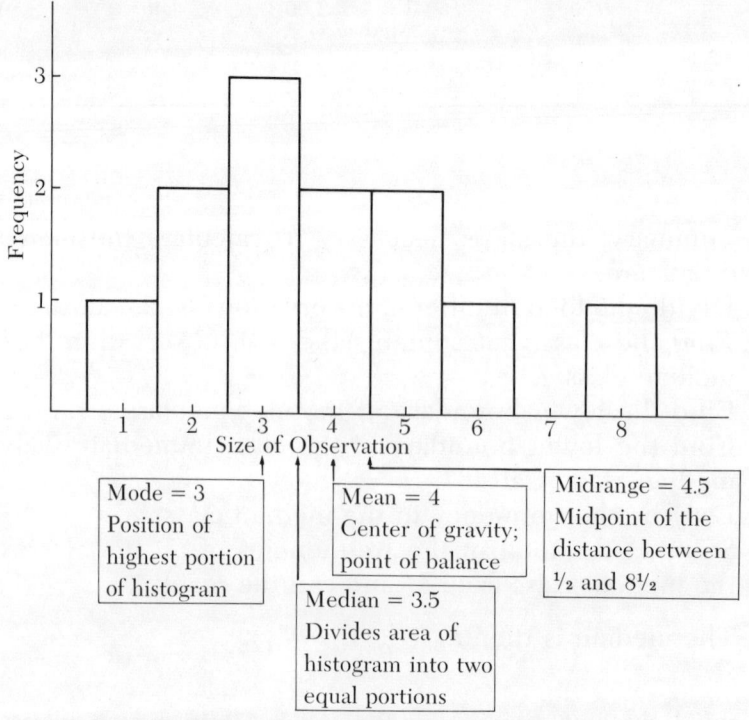

**Figure 3-1** Histogram of data showing the positions of the four basic measures of central tendency.

AVERAGES

EXAMPLE 3-8. A patient's temperature was taken 4 times one day and readings of 98.6, 103.2, 101, and 99.8 were recorded. The patient's "average" temperature for the day was $\frac{98.6 + 103.2}{2} = 100.9$, which is the midrange.

A graph quickly illustrates the differences between the various measures of central tendency. The following simple example is pictured by the histogram in Figure 3-1. Notice that the measures are designed to locate, in one way or another, the "center" of the distribution.

EXAMPLE 3-9. Consider the observations 1,2,2,3,3,3,4,4,5,5,6,8.

Mean = 4
Median = 3.5
Mode = 3
Midrange = 4.5

The "best" measure of location obviously differs according to what the experimenter wants to show. The measurement scale used also affects his choice of a location measure. For instance, nominal data merely classifies and has no numerical meaning (the "average" of a blond and a redhead is *not* a "bread-head"). Thus, for nominal data, only the mode can be used, because all other measures of location require some manner of numerical scale. Ordinal data does have some numerical meaning, although differences between ranks are usually meaningless. We employ the median in this case, since it divides the ordered observations into two equal groups; or in the case of grouped data, it divides the area of the histogram into two equal portions. Meaningful numerical measure results only from interval data, for which we use the mean, sometimes called the arithmetic average. We use the midrange for a quick measure, or if the physical situation is such that only minimum and maximum values can be recorded easily, such as with temperatures, stock prices, flood levels, and so on. Outliers have no effect on the mode or median, but they do affect the midrange drastically and can affect the mean seriously. Their presence may influence our choice of a measure of location.

EXAMPLE 3-10. The letter grades given on an English essay in a small senior class were C,D,A,B,C,E,A,C,B,D, and C. What was the "average" grade? These have at best ordinal measurement. First we shall arrange the scores: E,D,D,C,C,C,C,B,B,A,A. From this we notice that C is the middle grade or median. We could also give C as the mode or typical grade, since it occurs most often. Assume a grade range standard of: below 50 = E, 50–59 = D, 60–79 = C, 80–89 = B, and 90–100 = A. Suppose that we had available the original number scores from which the grades had been assigned; say they were, in ascending order, 26,51,55,67,69,73,78,82,89,91,92. Even though 8 out of the 11 grades are C or above, notice that the midrange $\left(\frac{26 + 92}{2} = 59\right)$ is in the D category, since it is highly influenced by

## AVERAGES

the low outlier of 26. The mean is 70.3. It is also affected somewhat by this outlier, though not so severely as the midrange. Among the numerical scores there is no mode.

**Exercise 3-1:** Given the observations 24, 10, 9, 4, 7, 9, 7, 3, 6, 7, find the mean, the median, the mode, and the midrange.

Mean     = _____

Median   = _____

Mode     = _____

Midrange = _____

Are they all the same number? _____
Which measure of location would be most appropriate for these data?
_____

Why? _____
_____

**Exercise 3-2:** Given the observations 1, 7, 3, 4, 3, 20, 2, 3, 4, 3, find the mean, the median, the mode, and the midrange.

Mean     = _____

Median   = _____

Mode     = _____

Midrange = _____

Are they all the same number? _____

Which measure of location would be most appropriate for these data?
_____

Why? _____
_____

**AVERAGES**

*Exercise 3-3:* Six students from a high school track team each entered in one main event at a state track meet. They placed as follows in their respective events: 5*th*, 2*nd*, 3*rd*, 1*st*, 2*nd*, and 10*th*. What is their "average" rank? _____ Which type of average is best to use in this case? _____ What objections might arise to using the midrange? _____

_____

*Exercise 3-4:* Five members of a high school home economics club entered a cake baking contest, a dress-making competition, a pickle-making contest, a quilting competition, and a flower-arranging competition at the state fair. They were awarded the places 3*rd*, 1*st*, 2*nd*, 2*nd*, and "no show" (meaning 11*th* place or below). What is their "average rank"? _____ Which type of average is best to use in this case? _____ What objections might arise to using the midrange? _____

_____

*Exercise 3-5:* Look around you at the nearest group of people and find the "average" hair color. _____

*Exercise 3-6:* Look around you at the nearest group of people and find the "average" sex. _____

*Exercise 3-7:* What kind of average would you consider best in the following instances? Why?

(a) Average heights of male athletes _____

_____

(b) Typical animal found in a zoo _____

_____

(c) Average heights of girls in a social club _____

_____

25

## AVERAGES

(d) Average salaries of all employees at the Ford Motor Company

(e) Average salaries of all assembly-line workers at the Ford Motor Company

(f) Average car driven by the employees of the Ford Motor Company

(g) Average price of General Motors Corporation stock on the New York stock exchange today

**Exercise 3-8:** What kind of average would you consider best in the following cases? Why?

(a) Average amount of rent charged in a college town for a two-bedroom apartment

(b) Average price charged for a ham sandwich in your hometown

(c) The typical flavor of milkshake ordered at a drive-in restaurant

(d) The average amount paid for breakfast by college sophomores

(e) The average length of time cars are parked at a certain parking meter in the center of town

(f) Average color of car driven by college presidents _____

_____

(g) Average temperature in your hometown yesterday _____

_____

EXAMPLE 3–11. Suppose the following numbers were written on separate slips of paper and put into a jar: 2,2,4,7,15. Imagine that we must predict the number that will be drawn by a person from this jar. The number we predict does not necessarily *have* to be one of the numbers on a slip of paper—we can predict *any* number. Each slip of paper has an equal chance of being drawn. What would be the best number to predict in the following cases:

(a) We get paid $15 for the prediction *if we are right*, and have to pay $20 penalty if we are wrong.

(b) We get paid $20 for the prediction regardless of whether it is right or wrong. However, we must pay an amount equal to the size of the error if we are wrong. In other words, if we predict "6", and "7" is drawn, we are paid $20 but have to pay back $1; or if we predict "10", and "2" is drawn, we are paid $20 but have to return $8.

(c) We get paid $25 for the prediction but must pay a penalty equal to the square of the size of the error if we are wrong. Thus, if we predict "6" and "7" is drawn we receive $25 and have to pay $1; if we predict "10" and "2" is drawn, we receive $25 and have to pay $64, since $8^2 = 64$.

For part (a) we would be foolish to predict a number not in the jar, and so we may restrict our attention to the numbers 2,4,7, and 15. If we predict "2", we have two chances out of five of being correct; but if we predict 4,7, or 15, we have only one chance in five of being correct. Therefore, the best prediction would be "2", which is the mode.

For part (b), it may be instructive to try a few values to see what happens. If we predict "2" this time, we will gain $20 for the prediction and lose nothing if a "2" comes up; we will lose $2 if the "4" is drawn; we will lose $5 if the "7" is drawn; and we will lose $13 if the "15" is drawn. Thus we would expect to lose $2 one-fifth of the time, $5 one-fifth of the time, $13 one-fifth of the time, and nothing two-fifths of the time. We can calculate the amount we would expect to gain or lose *on the average* by

$$\$20 - \left(\frac{2}{5}\right)(\$0) + \left(\frac{1}{5}\right)(\$2) + \left(\frac{1}{5}\right)(\$5) + \left(\frac{1}{5}\right)(\$13)$$

$$= \$20 - \left(\frac{1}{5}\right)(\$2 + \$5 + \$13)$$

$$= \$20 - \$4 = \$16 \text{ gain.}$$

## AVERAGES

*TABLE 3-2  GAINS OR LOSSES RESULTING WHEN VARIOUS NUMBERS ARE PREDICTED*

| Number Predicted | Number Drawn | Penalty (in dollars) | Expected Gain |
|---|---|---|---|
| 3 | 2<br>4<br>7<br>15 | 1 (for each "2")<br>1<br>4<br>12 | $\$20 - \frac{\$19}{5} = \$16.20$ |
| 4* | 2<br>4<br>7<br>15 | 2 (for each "2")<br>0<br>3<br>11 | $\$20 - \frac{\$18}{5} = \$16.40$ |
| 5 | 2<br>4<br>7<br>15 | 3 (for each "2")<br>1<br>2<br>10 | $\$20 - \frac{\$19}{5} = \$16.20$ |
| 6 | 2<br>4<br>7<br>15 | 4 (for each "2")<br>2<br>1<br>9 | $\$20 - \frac{\$20}{5} = \$16.00$ |
| 7* | 2<br>4<br>7<br>15 | 5 (for each "2")<br>3<br>0<br>8 | $\$20 - \frac{\$21}{5} = \$15.80$ |

*These values appear on numbered slips; the others do not.

In the same way (try it!) we see that if we predict the numbers 3,4,5,6, and 7, we can expect the various gains shown in Table 3-2.

Thus we see that we would expect to gain the most money on the average if we predicted the number "4", which is the median of 2,2,4,7,15. The median is clearly the best predictor in case (b).

**Exercise 3-9:** Set up a table similar to Table 3-2 to show that the mean is the best predictor for case (c).

***Exercise 3-10:*** Set up a table similar to Table 3-2 to verify that the mode is the best predictor for case (a).

The facts illustrated by Example 3-11 and by Exercises 3-9 and 3-10 are proved in detail in more advanced textbooks, and supply us with another method of determining which measure of central tendency would be most appropriate in certain kinds of experimental situations. If the experimenter wants to know the typical value, or the one single unique value that is most likely to occur, and if he is penalized for any wrong answer no matter how close or how far it may be from

# AVERAGES

the point of central tendency, then he should use the mode as his measure of central tendency (see Example 3-11, case (a)). If he needs the answer that is as close as possible to the true value and will be penalized only according to the size of his error, then he should use the median (see Example 3-11, case (b)). If, on the other hand, he knows that large errors are much worse than small errors, so that he will be penalized for a wrong answer in relationship to the square of the size of the error, then he should choose the mean (Example 3-11, case (c), illustrated by Exercise 3-9).

### TERMS TO REMEMBER

**Outliers:** Observations that are extremely large or extremely small in relation to most of the observations.

**Mean:** The arithmetic average, obtained by adding all the observations and then dividing the total by the number of observations.

**Median:** The value of the middle item, or the mean of the two middle items, when the observations are arranged in ascending numerical order. In the case of grouped data, the point dividing the area of the frequency histogram into two equal portions.

**Mode:** The most-often occurring observation in a group of observations.

**Midrange:** One-half the sum of the largest plus the smallest observation.

**Measures of Location or Central Tendency:** Kinds of averages; measures which indicate the point about which a distribution tends to center.

*chapter 4*

# SAMPLING FROM A POPULATION

Statistics has a language of its own. In this chapter we shall investigate "populations" that may not involve people, "samples" that are not left at our doors by salesmen, and "bias" that does not imply emotional prejudice. We shall initiate our investigation by imagining an experiment. Suppose we tossed two dice three times and observed the totals of the two numbers that were on the upturned faces of the dice each time. When the experiment is finished the data would consist of three numbers. The outcomes of our experiment might be as shown in Figure 4–1.

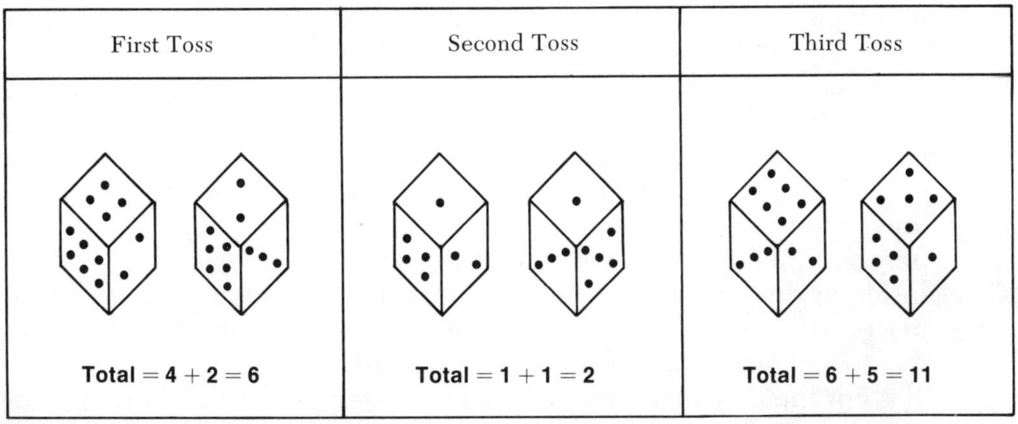

| First Toss | Second Toss | Third Toss |
|---|---|---|
| Total = 4 + 2 = 6 | Total = 1 + 1 = 2 | Total = 6 + 5 = 11 |

**Figure 4–1** One possible result of the experiment of tossing two dice three times and recording the totals.

31

## SAMPLING FROM A POPULATION

In performing this experiment and gathering our data, we have sampled from a population and observed events from a sample space. The population consists of all possible sets of two upturned faces on dice, and the sample space includes all possible totals of the two numbers showing on these upturned faces. Thus, by "population" we do not necessarily mean a group of people. In fact, we often change the definition of our statistical population according to the context of the problem. For example, all students currently enrolled at a certain school would be the population considered if we wished to study the current marital status of these students. These same students would constitute a sample if we wished to study the current marital status of all students in the United States. They would comprise the most recent sample if we wished to compare the marital status of students now at that school with that of students who had attended the school in years past.

For the dice experiment above, we can specify all the items of the population. It has 36 items, the sides having on them, respectively, the numbers (1,1), (1,2), (1,3), (1,4), (1,5), (1,6), (2,1), (2,2), (2,3), (2,4), (2,5), (2,6), (3,1), (3,2), (3,3), (3,4), (3,5), (3,6), (4,1), (4,2), (4,3), (4,4), (4,5), (4,6), (5,1), (5,2), (5,3), (5,4), (5,5), (5,6), (6,1), (6,2), (6,3), (6,4), (6,5), (6,6). Any **population** consists of all possible items concerning which measurements could be taken, or the total number of potentially observable units. A **sample** is a subset of the population—a set of items chosen from the population and then observed and measured by the experimenter. Our sample, three individual items from the population, yielded the observations (4,2), (1,1), and (6,5).

Whenever a sample item, once drawn, is replaced so that it can recur in the sample, the sample is said to have been drawn **with replacement.** For example, in tossing dice, after observing a "1" on one toss, it is possible to observe a "1" also on a later toss. If, on the other hand, the experimenter removes an item from the population after observing it, then he is said to be **sampling without replacement.** An example of this type of sampling occurs if a child draws a handful of jellybeans from a bag, observes that he has chosen 2 black ones, 1 red one, and 2 white ones, and then promptly eats them. Those jellybeans cannot be drawn in a later sample. If we were drawing a sample of size 10 from an urn containing colored beads, we might sample the colors *without replacement* by merely removing ten beads at one time and observing the distribution of their colors. Alternatively, we may draw the sample *with replacement* by drawing one bead at a time, observing its color, and placing it back in the urn before stirring the beads to draw again. Because the sampling of dice is done with replacement, observing six tosses of one die is equivalent to observing one toss of six dice, so long as the dice are all fair.

For the dice experiment above we observed the numbers appearing on the upturned faces of the dice in order to obtain their totals. All possible values of these totals—namely 2, 3, 4, 5, 6, 7, 8, 9, 10, 11, and

12—constitute the sample space of the experiment. Thus, from the population, we may define a **sample space** as the set of all possible measurements that could be taken on a population, or the set of all possible individual outcomes of an experiment. The terms population and sample space may seem almost interchangeable. The basic distinction between them is that the population contains the items or things observed by the experimenter (such as the sides of a die, which can be identified by the numbers thereon), while the sample space consists of all possible measurements that could be taken (such as 1, 2, 3, 4, 5, or 6). The three events which we observed in our experiment were the events 6, 2, and 11. An **event** is merely one or more items of the sample space, or, in other words, any collection of individual items. The terms "event" and "outcome" are interchangeable. Next we shall consider ways of determining how likely it is that a given event will occur.

We often contemplate a team's chances of winning, the chance a person has to pass a course, the chances of rain today, and so on. The common usage of the word "chance" involves nothing more than converting a probability to a percentage statement. The **probability** of an event is the proportion of the time we would expect that event to occur in the long run. Probabilities of events are numbers that range from 0 to 1, which tell what fraction of a large number of trials the desired event will occur. The notation used to express a probability is $P(\text{event}) = \text{number}$. If A is some event and $P(A) = 0.2$, then the probability that A will happen is two-tenths or, in other words, there is a 20% chance that A will occur.

Let us evaluate the chances of the events in the dice experiment. To do so we may wish to calculate the probability of each of the three events, getting a "2", getting a "6", and getting an "11". From the population of 36 equally likely items we see that only one, the item (1,1) would yield a total of "2", or in other words, $P(\text{a total of "2"}) = 1/36$. There are five ways to get a total of "6", namely (1,5), (2,4), (3,3), (4,2), and (5,1); so we would say $P(\text{a total of "6"}) = 5/36$. In a similar manner we would assign a probability of 2/36 to the event of "getting a total of 11". If we wished, we could assign a probability of 12/36 to the event "getting a total that is divisible by 3" which is the same event as "getting a total of 3 or 6 or 9 or 12".

The probability statements we hear should be more understandable to us now. If the weatherman tells us there is a 30% chance of rain today, he means that if the particular weather conditions prevalent today were repeated many, many times, then on 30% of those days it would rain. If a toothpaste manufacturer claims that we may expect 32% fewer cavities with his toothpaste, then supposedly if thousands of people brushed (under carefully controlled conditions) with his toothpaste they would, on the average, experience 32% fewer cavities than did those unfortunate people who brushed with one of the other

## SAMPLING FROM A POPULATION

toothpastes used by his technicians in performing the laboratory test of his toothpaste. Sometimes it is difficult to conceive what is really meant by "the proportion of the time we would expect the event to occur in the long run." Suppose, for example, that a politician claims that he has a 70% chance of winning an election in his state. He will only run for this election once, and he will either win or lose the election. Still, the same idea is meant to be conveyed by stating that the probability of his winning the election is 0.70—namely, if the same conditions were to be present many times, then in 70% of these cases he would win the election.

In calculating the probabilities of our events, we mentioned that each item in the population was equally likely; this, however, is true only if fair dice are tossed fairly. If we toss the dice in a fair manner, such as throwing them hard against a vertical surface, then our sample can be considered a random one. A **random sample** (sometimes called a "simple random sample") is a sample of items which is chosen in such a way that any item in the population has an equal opportunity of being included in the sample. An experimenter always hopes his sample will be representative of the population. A random sample is more likely to reflect the true nature of the population than one chosen in any other way.

In our experiment with the two dice, the **sampled population,** the set of items which we actually did sample, coincided with our **target population,** the set of items which we specified that we wanted to sample. Often in practice, however, the sampled population is different from the target population. In other words, the experimenter wants to investigate a certain set of items—his target population; this may or may not be the group which is actually sampled—his sampled population. From his experimental data, he must be careful to draw inferences only about the sampled population if it differs from the target population. For example, suppose we conduct a poll of voters prior to an election to see which presidential candidate they favor. The target population would be all the people who actually will vote in November. But how can we get a list of their names two months before November? It is impossible to get this list; so we must content ourselves with using a list of the currently registered voters, the sampled population. We hope the sampled population will be the same as the target population. Obviously, however, it may not be the same, since not all registered voters will vote. The following three examples summarize these concepts.

EXAMPLE 4–1. A poll is to be conducted at 6:45 P.M. to determine how many people are watching a news program on TV. The *target population* is the set of people who own TV sets. Since a telephone survey is the only feasible way to obtain the information, the *sampled population* is the set of people who have telephones. Suppose we called all people who have a "1" as the last digit of their phone numbers; this procedure would yield a *simple random sample* consisting of 10% of the sampled population. If we called all the people who have a "13" as the last two digits, then we would have a 1% *simple random sample*. Ignoring unanswered calls and busy signals, we could identify the points of the *sample space* by assigning "0" to the answer "no" and "1" to the answer "yes". When an individual is called and asked whether or not he is watching a TV news program, an answer of "yes", recorded as a "1", would be an *event*. To illustrate further, let's change the sample space by a slight variation of procedure: suppose we wish to record whether the person called is watching ABC, NBC, CBS, or another network, or not watching TV at all. Then we may wish to have a sample space such as 0 = not watching TV; 1 = watching NBC; 2 = watching CBS; 3 = watching ABC; 4 = watching another network. Because the same person would not be interviewed twice, this experiment is an example of *sampling without replacement*. If a person responded that he was not watching TV at all, the answer of "no", recorded as "0", would be an *event*. The *sample space* in this case consists of the numbers 0, 1, 2, 3, and 4.

EXAMPLE 4-2. Suppose we want to learn the average number of hours per week each person spends in the local library. The *target population* consists of all people who use the library. But the *sampled population* might be composed of those who have library cards and who have not moved away since they got their cards. The record of library cards issued may comprise the only listing of library users available. A *simple random sample* could be obtained by mixing the cards thoroughly and then choosing the people listed on the top 10 cards (or 15, or 50, according to how large a sample is desired). Those chosen would then be phoned or written or interviewed personally to learn how many hours per week each spends in the library. The *sample space* could contain any number from 0 to the total number of hours per week that the library is open. An *event* could be the number "3.5" recorded when an individual contacted responded that he spent about 3.5 hours per week in the library. Again, in this example, we are *sampling without replacement*.

EXAMPLE 4-3. When we toss a single die[*], the *population* consists of the six possible upturned sides. There is no distinction in this experiment between the sampled and the target populations — unless some of the spots have been rubbed off one side of the die or unless it is a trick die that does not have all the numbers on it. The *sample space* consists of the numbers 1, 2, 3, 4, 5, 6. A *simple random sample* (of size 1) is obtained when we toss the die hard against a vertical surface, allowing it to hit and fall (see Figure 4-2) and then observe the number on the upturned face. If we were to hold the die rigidly be-

---

[*]A die is a single one of a pair of dice.

## SAMPLING FROM A POPULATION

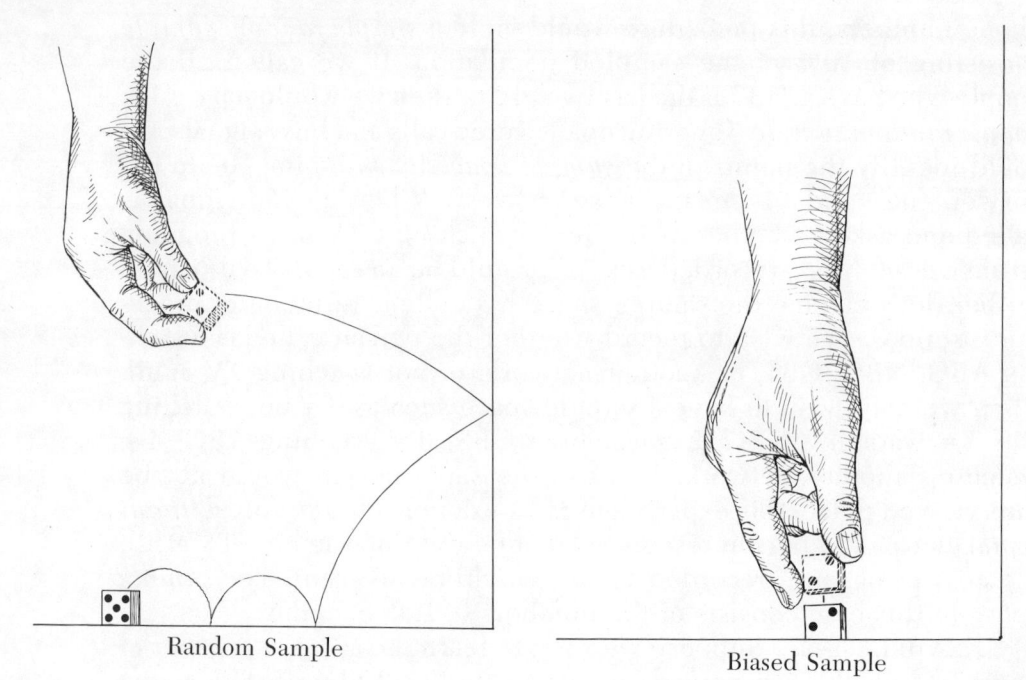

**Figure 4-2** Illustration of random and biased samples.

tween thumb and forefinger about one quarter inch above the surface of the table and then drop it carefully to avoid having it turn or roll at all, we would have a **biased sample**. This type of nonrandom sample results when certain items of the population have greater chances of occurring in the sample than do other items. One possible *event* would be getting a number less than 3. The *probability* of this event would be 2/6, because this event occurs when either a "1" or a "2", out of the six equally likely numbers, turns up. If the die is only tossed once, there is no question of sampling with or without replacement.

Notice that in Examples 4–1 and 4–2 we could not calculate the probabilities of possible events since we had no "past experience" or "model" to call upon, while in Example 4–3, assuming a fair die, we would expect each of the six possible points in the sample space to be equally likely. Thus we know what probabilities to assign to each possible event. A **model** gives a mathematical rule, physical law, or general principle upon which to base the calculation of probabilities of events for any given experimental situation. The model idealizes the physical situation and thereby specifies the possible events that can occur. Sometimes the purpose of sampling is to see whether our experimental situation fits a certain model. For instance, in Example 4–3 we may be interested in tossing the die several times to learn whether the sample data indicate that the die is "fair."

Sometimes, such as in the situations outlined in Examples 4–1 and

4–2, we may be gathering the data for the purpose of determining what kind of model might apply to that particular situation: advertisers may question the probability of reaching a certain group of people by advertising with the evening news, or the administrators of the library may need to know what proportion of the library users would be affected by a proposed change in library hours. A knowledge of how to calculate and work with probabilities of events and relate them to the models involved is basic to their use in statistical analyses.

EXAMPLE 4–4. Suppose that on a table next to you lies a five-pound box of candy containing 50 chocolates in the top layer. Suppose that you and your little brother both like chocolate-covered cherries best. Your little brother is not tall enough to see what he is choosing but can reach any piece in the box. You are reading a book and do not bother to observe which piece you are choosing. Thus you both could be assumed to be choosing a *simple random sample.* Your little brother's procedure is to choose a piece of candy, stick his thumb into it, and, if it is not a chocolate-covered cherry, replace it. Thus he is *sampling with replacement.* Suppose that after ten trials of this nature, he has found only one chocolate-covered cherry, and has eaten it. Then, based on his sample data, chocolate-covered cherries constitute 10% of the pieces in the top layer. His sample estimate for the probability of finding a chocolate-covered cherry is 0.1. In this type of example, we are using sample evidence to infer results about the population (of chocolates in the top layer of the box of candy). Perhaps you, being more experienced, know that in the top layer there are ten chocolate-covered cherries. In other words, you know the original model involved, namely $P(\text{chocolate-covered cherries}) = \frac{10}{50} = 0.2$.

Also assume that you are less finicky and eat whatever you choose; thus you are *sampling without replacement.* Because your brother has already eaten one cherry, there are 49 pieces of candy left, of which only 9 are cherries. Therefore, for your first choice, $P(\text{chocolate-covered cherries}) = \frac{9}{49}$. Incidentally, the probability that on your first choice you get a piece of candy with a thumb imprint in it is also $\frac{9}{49}$, since little brother stuck his thumb in ten pieces and only ate one. If your first choice is not a cherry, then on the second draw, the probability of getting a cherry is $\frac{9}{48}$. If the first four choices are all not cherries, then the probability of getting a cherry on the next draw is $\frac{9}{45}$. Suppose that you have chosen ten pieces of candy and found 3 cherries. Your sample estimate of $P(\text{chocolate-covered cherry})$ would then be $\frac{3}{10}$, and $P(\text{chocolate-covered cherry}) = \frac{6}{39}$ when your brother begins sampling again.

## SAMPLING FROM A POPULATION

Notice that in Examples 4–1, 4–3, and 4–4 our possible observations are very few and thus cannot differ much from one another. In the first situation of Example 4–1 there are only two possible observations (0 and 1); in Example 4–3, there are six (1,2,3,4,5, or 6); and in Example 4–4 there can be at most eleven (0,1,2, etc. indicating the number of cherries eaten). In Example 4–2, however, the numbers could vary quite widely, ranging from 0 to the total number of hours per week that the library is open.

If a population contains observations that are quite variable, then the experimenter needs a larger sample to describe the properties of the population. The basic purpose of sampling is to infer the properties of the population based on the sample characteristics. For instance, we can use the mean of a random sample from a population as a rough approximation of the mean of the population itself. These estimates are based on sample information. Their accuracy in describing the true characteristics of the population depends basically upon two things: how diverse the items drawn tend to be, and how large a sample is taken. If the items in one population are less diverse (less varied, more similar to each other) than those of a second population, then for samples of a given size, the sample-based estimates of the mean of the first population will usually, though not always, be more accurate than the estimates of the mean of the second population.

For example, consider two populations which both yield observations that have mean equal to 5:

Population A: 5.00, 5.02, 5.01, 5.00, 4.98, 5.00, 5.01, 4.98, 4.97, 5.03

Population B: 1, 7, 5, 3, 1, 10, 6, 2, 10

Consider now all possible samples of size two that could be drawn without replacement from these populations. What would be all of the possible sample-based means that could result? The sample size is two, and the two smallest values in population A are 4.97 and 4.98, while the two largest are 5.02 and 5.03. Thus from population A, the smallest possible sample mean would be $(4.97 + 4.98)/2 = 4.975$, and the largest would be $(5.02 + 5.03)/2 = 5.025$. Both the smallest and the largest sample means possible are still very close to the population mean, which is equal to 5. In population B the smallest possible sample mean would be $(1+1)/2 = 1$ and the largest would be $\frac{10+10}{2} = 10$, neither of which is very close to 5. However, if samples of size 6 were taken without replacement, then for population A the smallest possible value would be

$$\frac{4.97 + 2(4.98) + 3(5.00)}{6} = 4.988$$

and the largest would be

$$\frac{5.03 + 5.02 + 2(5.01) + 2(5.00)}{6} = 5.012,$$

both of which are closer to 5 than were the estimates of 4.975 and 5.025 obtained before. For population B, the smallest possible value would be

$$\frac{1 + 1 + 2 + 3 + 5 + 6}{6} = 3.000$$

and the largest would be

$$\frac{10 + 10 + 7 + 6 + 5 + 3}{6} = 6.833,$$

which also are closer to 5 than the previous estimates of 1 and 10. By comparing the possible means of these two samples, it becomes obvious that the precision of the sample estimate depends on both the variability in the data and the number of items included in the sample. Samples of larger sizes and samples of less variable measurements usually yield more precise sample estimates of population characteristics. This rule also holds when items are replaced after sampling so that they can be drawn again and repeated observations on the same item may result.

Often the experimenter wishes to estimate the expected proportion of items having a given attribute in a set of similar items. For example, we may wish to estimate the chance of getting "heads" with a fair coin. In this case, the sample proportions will tend to get closer and closer to the true probability of obtaining that attribute—in other words, closer to the model—as the sample size gets larger. If we count the number of heads when a coin is tossed 10 times, then $p_1 = \frac{\text{the number of heads}}{10}$; when tossed 100 times, $p_2 = \frac{\text{the number of heads}}{100}$; when tossed 1000 times, $p_3 = \frac{\text{the number of heads}}{1000}$; and so on. The value $p_3$ would be closer, usually, to the model that states $P(\text{heads}) = 1/2$, than either $p_2$ or $p_1$. This principle is called the **Law of Large Numbers.** Notice that we have *no guarantee* that a larger sample size will yield more precise estimates of $P$. Instead we say "this is *usually* the case" or "the sample proportion will *tend* to get closer to the model's probability" or "$p_3$ would be closer, *usually*, to the true

## SAMPLING FROM A POPULATION

value." In observing chance events we may never be certain of the outcome but can only predict what we *expect* to occur *in the long run*. In other words, we may expect sampling errors to occur, but in the long run they tend to "even out" or else be outweighed by other observations. Two more heads than expected in 10 tosses of a coin is a lot, but two more heads than expected in 10,000 tosses is negligible.

How do we decide on the model in the situation just described? It was a fair coin, so the model $P(\text{heads}) = .5$ specified the probabilities of all possible outcomes. Merely saying that a coin or die is fair tells us that we may assume that all of the possible events—"heads" or "tails" for the coin and 1, 2, 3, 4, 5, or 6 for the die—are equally likely; thus we may assign equal probabilities of 1/2 and 1/6 respectively to their outcomes.

Suppose we have a jar which contains 50 pennies, 25 nickels, 30 dimes, 15 quarters, and 5 half dollars. Then the probabilities of drawing a stated kind of coin at random from the jar are $P(\text{penny}) = \frac{50}{125}$; $P(\text{nickel}) = \frac{25}{125}$; $P(\text{dime}) = \frac{30}{125}$; $P(\text{quarter}) = \frac{15}{125}$; and $P(\text{half dollar}) = \frac{5}{125}$. Note first that the sum of all of these probabilities is $\frac{125}{125} = 1$; that is, it is certain that we will draw a coin. Second, note that, had we used beads of various colors rather than coins (as long as there were 50 of one color, 25 of a second color, 30 of a third color, 15 of another, and 5 of another), the numerical probabilities would have remained the same; that is, the *model* is the same for both experiments.

***Exercise 4-1:*** Fill a bucket or jar with at least 50 balls, beads, or marbles of different colors. In Table 4-1, list the colors represented and the true probability of drawing each color. You may obtain the true probability for each color by dividing the number of balls of that color by the total number of balls in the jar. Before drawing each sample, mix the balls thoroughly. Draw a sample of size 1 and note its color. In column 1 under "Samples of Size 1," find the row for that color and mark a "1" there; put zeros in all the other rows to indicate that no balls of other colors were drawn in that sample. Return the ball to the jar, mix the balls, and draw another sample of size 1. Note its color in column 2. Repeat this procedure until you have drawn five samples of size 1.

Now draw a sample, without replacement, of size 10. Note the number of balls of each color in the sample; mark them in the appropriate rows in the table, in column 1 under "Samples of Size 10." Replace the balls, mix, and draw another sample of size 10. Note the frequencies of the colors in column 2. Repeat this procedure until you have drawn five samples of size 10.

Finally, in the same manner draw five samples of size 25 and note

**TABLE 4-1** SAMPLING TABLE

| Color | Probability of the Color | Samples of Size 1 ||||| Samples of Size 10 ||||| Samples of Size 25 |||||
|---|---|---|---|---|---|---|---|---|---|---|---|---|---|---|---|---|
| | | 1 | 2 | 3 | 4 | 5 | 1 | 2 | 3 | 4 | 5 | 1 | 2 | 3 | 4 | 5 |
| 1. | | | | | | | | | | | | | | | | |
| 2. | | | | | | | | | | | | | | | | |
| 3. | | | | | | | | | | | | | | | | |
| 4. | | | | | | | | | | | | | | | | |
| 5. | | | | | | | | | | | | | | | | |
| 6. | | | | | | | | | | | | | | | | |
| 7. | | | | | | | | | | | | | | | | |
| 8. | | | | | | | | | | | | | | | | |
| 9. | | | | | | | | | | | | | | | | |
| 10. | | | | | | | | | | | | | | | | |

## SAMPLING FROM A POPULATION

the frequencies in the table. Note that each sample is drawn *without replacement;* each sample is then replaced before another sample is taken.

**Exercise 4–2:** (a) Were the samples chosen in Exercise 4–1 random? _____ Why or why not? _____

_____

(b) Would you expect the probabilities based on samples of size 25 to conform to the probabilities based on theory (to the model) more than those for samples of size 1 or size 10?

_____ Why or why not? _____

_____

(c) Combine the results of all five samples of size 25 and estimate the true probability of each color based on your combined sample results. In other words, take the five numbers for each color in the "Samples of Size 25" column, add them together, and then divide by 125 to get the average number of balls observed of each color. Using this combined sample of 125, along with the first sample of each sample size, fill in Table 4–2 and discuss whether you believe that your results are unusual. For example, if in your first sample of size 1, you observed a black ball, then you would fill in a "1" in the row labeled "black" and the column labeled "1" and "0's" elsewhere in that column. If in the first sample of size ten you observed 2 black, 1 red, 5 white, and 3 green, you would put .2, .1, .5, and .3 in the appropriate rows under the column labeled "10." If in the first sample of size 25 you observed 3 black, 1 red, 13 white, 5 green, 1 pink, and 2 blue, you would put .12, .04, .52, .20, .04, and .08 respectively for each color. Finally, if in the other four samples of 25 you observed 2, 4, 3, and 1 black balls respectively, you would put $\frac{3+2+4+3+1}{125} = .104$ in the row labeled "black" in the column labeled "125" and do the same for each other color. Two or three decimal places of accuracy is sufficient for the values in this table. Notice that the combined sample of 125 could not be classified either as a sample with replacement or as one without replacement, since each sample of size 25 was drawn without replacement, but all 25 items were replaced before the next sample of 25 was drawn.

**SAMPLING FROM A POPULATION**

**TABLE 4-2** PROBABILITY TABLE

| Color | Prob. (Color) (Based on Theory) | Prob. (Color) (Based on Sample Evidence) |||| 
|---|---|---|---|---|---|
| | | 1 | 10 | 25 | 125 |
| 1. | | | | | |
| 2. | | | | | |
| 3. | | | | | |
| 4. | | | | | |
| 5. | | | | | |
| 6. | | | | | |
| 7. | | | | | |
| 8. | | | | | |
| 9. | | | | | |
| 10. | | | | | |

Discussion: _____

_____

_____

(d) What population was sampled in Exercise 4–1? _____

_____

(e) Was the sampled population the same as the target population?

_____

(f) Name three events which could have occurred when you drew a sample of size 1. _____  _____  _____

**TERMS TO REMEMBER**

**Population:** All possible items concerning which measurements could be taken; or the total aggregate of potential units for observation.

**Sample:** A subset of the population; the set of items chosen from the population and then observed and measured by the experimenter.

43

## SAMPLING FROM A POPULATION

**Sampling With Replacement:** Observing items in such a way that each item is replaced after recording its measurement so that it may be chosen for observation again in the same sample.

**Sampling Without Replacement:** Observing items in such a way that the item is removed from the population and is no longer available for observation in the same sample.

**Sample Space:** All possible measurements that could be taken on a population; in other words, the set of all possible individual outcomes of an experiment.

**Event:** Any collection of individual outcomes; one or more of the items of a sample space.

**Probability:** The proportion of the time that one would expect that an event will occur in the long run.

**Simple Random Sample:** A sample of items which is chosen so that all items in the population have an equal opportunity of being included in the sample.

**Sampled Population:** The set of items that the experimenter actually *does* sample.

**Target Population:** The set of items that the experimenter *desires* to sample.

**Biased Sample:** A sample of items chosen in such a way that one (or more) of the items of the population has a greater chance of occurring in the sample than do other items.

**Model:** An idealization of a physical situation by means of which one may calculate the probabilities of all possible events that could occur for that situation.

**Law of Large Numbers:** As the sample size grows larger and larger, the proportion of items having a given attribute that is observed in data from a random sample gets closer and closer to the true proportion of items having that attribute in the sampled population.

*chapter 5*

# THE CONCEPT OF RANDOMNESS

When we drew samples of balls from an urn in Chapter 4, the balls were first mixed well and then drawn without regard to which ones were chosen for each sample. By doing so, we were assured that each ball had an equal chance of being chosen for the sample. Thus, by the definition of Chapter 4, we had a simple random sample. When drawing a random sample with replacement from a population of size $n$, each ball in the population has a probability of $\frac{1}{n}$ of being chosen. If we draw a random sample without replacement, each ball has a probability of $\frac{1}{n}$ of being chosen first; each of those left has a $\frac{1}{n-1}$ probability of being chosen second; each of those left then has a $\frac{1}{n-2}$ chance of being chosen third, and so on.

It is easy to stir balls in an urn to assure a random choice. But people, or rats, or pea plants, cannot be dumped into a pot and stirred;

## THE CONCEPT OF RANDOMNESS

in these cases we need some other procedure for choosing a simple random sample. **Tables of random digits** are available in many mathematical table books. An example is Table I of the appendix. Reading from left to right, or from top to bottom (or in any systematic manner) we find that the probability of finding any single digit (0, 1, 2, 3, 4, 5, 6, 7, 8, or 9) in any particular place is approximately 1/10; for this reason they are called random digits. Numbers in random digit tables are sometimes said to be in a **random sequence,** or in **random order,** since it is impossible to predict that any particular number will occur in a certain place by knowing the predecessors and/or the successors of that number. Suppose that we wanted to choose a single item randomly from a group of ten items. We could number the items from "0" through "9" and then decide to select, say, the first digit in the third column of numbers in Table I, without knowledge beforehand of what number is there. Because it happens to be a "2", we would then choose the second item. That item would be called our "random choice," and our procedure has resulted in a "random sample" of size 1. We could state that we have employed an "element of randomness" in choosing the item labeled "2".

EXAMPLE 5–1. Suppose that an instructor must choose 7 students from a class of 36 to participate in a TV news program. To avoid showing partiality, he decides to choose the 7 students randomly. First he assigns each student a number between 1 and 36. He might then employ one of two different methods to choose his group randomly.

*Method 1:* He could begin at any point in Table I and read the numbers from left to right in a row using two-digit groups and ignoring all numbers except 01 through 36. Say that he began with the second row and seventh digit. Then the sequence of numbers would be 6 6 2̲1̲ 6 5 2̲5̲ 2 5 4 8 2 5 3̲2̲ 6 6 8 9 2 1 7 4 2̲3̲ 1 5̲2̲ 6 6 2̲0̲8̲ and he would choose the seven students having the underlined numbers as his group. Notice that "25" occurs three times and "21" twice. He would ignore these repeated occurrences and record only one "25" and one "21".

*Method 2:* Alternatively, he could begin at any point in Table I and read down the column, choosing only the first two columns of digits from each group. If he began with the top number of the third group of numbers, his sequence would read: 2̲9̲, 65, 1̲3̲, 1̲5̲, 2̲8̲, 73, 99, 78, 90, 57, 76, 2̲7̲, 52, 2̲1̲, 39, 92, 1̲1̲. Thus he would choose the students with the numbers 29, 13, 15, 28, 27, 21, 11. Many other methods could easily be designed to enable him to make a random choice. If a table such as Table I is not available, the experimenter may open the phone book to any page and use the last 3 digits of the phone numbers as a random number table (the last 4 digits are random only in a large city phone book).

It is important to know how to draw random samples. Are the

## THE CONCEPT OF RANDOMNESS

following examples really methods of drawing random samples? If so, then any observation must have an equal chance of being chosen.

**Exercise 5-1:** In a group of 50 people we wish to draw a sample of size 25. We line the people up and as each person steps forward, we toss a coin. If it lands heads, the person is accepted for the sample; if it lands tails, he is rejected.

(a) Would the sample be randomly drawn? _____

(b) Would the sample contain exactly 25 members? _____

If not, how could you modify the procedure so that 25 members would be randomly chosen? _____

_____

**Exercise 5-2:** Out of a group of 6 rats, we want to choose 2 for a psychological experiment. We number the rats 1 through 6 and toss two dice. We choose the 2 rats whose numbers show on the upturned faces of the dice.

(a) Is the sample randomly drawn? _____

(b) Does it contain exactly 2 elements? _____ If not, how could the procedure be modified so that exactly 2 rats would be randomly drawn? _____

_____

When we draw a sample of items for an experiment, we often intend to subject different ones of the items to different experimental treatments. Besides insuring a random choice of items, we must also be sure that the selection of any particular item for any particular treatment is also random. This restriction usually presents no basic problem, since if items are chosen in a random sequence, then their assignment to the different treatment groups will also be random.

EXAMPLE 5-2. Suppose that from a flat of 69 tomato plants in a greenhouse, an experimenter wants to choose 15 plants and subject 5 of them to treatment A, 5 to treatment B, and 5 to treatment C. He could then number the plants in any sequence and use Table I to select the 15 plants in a manner similar to Example 5-1, Methods 1 or 2.

## THE CONCEPT OF RANDOMNESS

He would put the first 5 aside for treatment A, the next 5 for treatment B, and the last 5 chosen for treatment C. In this manner, the 3 sets of 5 plants each would be chosen randomly from the 69 available plants and assigned randomly to the treatments.

***Exercise 5-3:*** The last four digits of the numbers in a phone book occur in random order in large cities (in small towns only the last 3 digits are random). How could you use these "random digits" to choose two samples of 25 people each from a student roll of 6000 students in a small university? _____

_____

_____

_____

***Exercise 5-4:*** Suppose that an experimenter wants to learn whether certain chemical weed control treatments affect germination of corn seeds. He has a bag of thousands of corn seeds from which to choose a sample of size 100. He wants to assign 20 seeds randomly to each of 5 separate chemical treatments. How could a random numbers table be used to choose the random sample and assign the seeds to the 5 treatments? _____

_____

_____

_____

***Exercise 5-5:*** A simple sampling device can be made by numbering 10 discs or beads or ping-pong balls with the digits from 0 to 9, and placing them in a box or urn or bucket. If the objects are mixed and one is drawn, then the number so drawn has been randomly chosen. How could one use this kind of sampling device to assign randomly 20 rats to 4 different experimental conditions? _____

_____

_____

_____

## THE CONCEPT OF RANDOMNESS

**Exercise 5-6:** A circular spinner may be marked off into 10 equal-area portions, labeled 1, 2, 3, 4, 5, 6, 7, 8, 9, and 10. How could this be used to assign 24 people randomly to 3 classrooms in such a way that each classroom would have exactly 8 people? _____

_____

_____

_____

**TERMS TO REMEMBER**

**Random Digits Table:**  A table listing the digits 0, 1, 2, 3, 4, 5, 6, 7, 8, 9 in such a way that the probability of any digit occurring in any particular position is 1/10.

**Random Sequence:**  A sequence of observations in which it is impossible to predict any single value given knowledge of its predecessors or its successors.

**Random Order:**  The positions of items in a random sequence.

chapter 6

# A ONE-SAMPLE TEST OF RANDOMNESS

Often the experimenter must determine whether observations made sequentially in a certain time period or in a certain location are in random order. He may be investigating a learning experience. For example, typewriting skills should improve with practice, so that successive speed and accuracy scores should increase. Also, a child's ability to read should increase as he is promoted from one grade to the next. The sequence of courses taken in mathematics is quite important since each mathematics course has concepts which require a previous knowledge of mathematics. If, for any of these cases, a random sequence of better or worse scores is exhibited, then learning has not occurred. In this chapter we shall be concerned with observing sequences in which the items can be classified into one of two categories like "success" or "failure", "high" or "low", "male" or "female", "better" or "worse", and so forth.

Consider a line of children waiting to get a drink of water at the water fountain in a grade school hallway. Will the boys tend to stand in one "bunch" and girls in another, or will the different sexes be interspersed randomly in the line? In the language of statistics, will the sequence consist of two runs? By the term **sequence** we mean items which have a certain ordering in time or space. A **run** is a sequence of objects of one type that is preceded and succeeded by either no objects at all or else by objects of another type.

## A ONE-SAMPLE TEST OF RANDOMNESS

Consider a sequence of successive tests showing successes and failures in a missile-testing project. Suppose that the time sequence of successes and failures was observed to be: F F F F S S F S S S S S S S S S S S. This sequence has four runs. If we observe only the first six tests, F F F F S S, it is difficult to "predict" the result of the seventh test. However, after nineteen tests, we would definitely expect a success on the twentieth trial because it is quite obvious that a pattern of consecutive successes has been established and that the S's and F's are no longer occurring randomly. The director of the project would be quite concerned with the pattern of the sequence of successes and failures for the tests. Either randomness in the sequence or else a sequence of failures would imply that the missile testing should be discontinued. The sequence should have a "bunching" of successes at the latter end if the missile is being perfected. This bunching of successes — or of failures — is called **clumping**. A tendency toward very long runs, or clumping, implies a non-random sequence of events.

To illustrate further, the quality control supervisor on an assembly line at a factory expects a random sequence of good and defective products if the production line is in control. Some defectives will always occur randomly in assembly line manufacture, but they should not occur consistently and sequentially. If there is a machine or an operator malfunctioning, a clumping of defective products will occur. As soon as his testing uncovers this long run of defectives, the quality control supervisor immediately stops the production process until the defective machine is adjusted or the operator is instructed. Thus a measure of the "clumpiness" of the observations in terms of the number of runs observed often indicates whether the observations are in random order. Too few runs usually indicates a lack of randomness, although few runs could occur by chance in random sampling.

Furthermore, too many runs may indicate a lack of randomness. In tossing a coin ten times you would be just as suspicious of non-randomness if you obtained HTHTHTHTHT as if you obtained all H or all T. You may suspect that the person tossing the coin can control the outcome. But how many is "too many" and how few is "too few?" The answer to this question often differs according to the experimental situation. If a right or wrong answer meant that we would gain or lose a dime we might not be too concerned about how hard we worked to find the answer; but if a right answer meant we would gain $100 and a wrong answer that we would lose $100, we would be more interested in being "sure" before answering.

As we have said before, whenever we gather experimental data, we are sampling in order to learn about a population. From a representative sample, we may expect to get a true picture of the population characteristics with only a few observations. On the other hand, if there happen to be some "odd" or outlying observations in the sample, a larger number of observations would be needed to offset their effects. Unfortunately, when we draw a sample in practice, we usually cannot

## A ONE-SAMPLE TEST OF RANDOMNESS

tell by observing the values whether they really yield a representative sample—we can only hope they do.

To illustrate this problem we shall reconsider the two populations, A and B, from Chapter 4. Population B had the measures 1, 7, 5, 3, 1, 10, 6, 2, 10. Suppose that we had drawn a random sample of size two from population B and just happened to observe the two 1's. Without knowledge of the population measurements, we would not know that this sample is not representative. In practice, we do not know the population values; without them, we have no way of knowing whether or not the sample is representative.

In a similar manner, if the number of runs we observe in a certain sample of values is such that that number of runs has a probability of less than 10% of occurring in a random sequence, then either the sequence is not random or we have observed an event having small probability. Either explanation is possible and we do not know which is true, but we would feel that there is good evidence that the sequence is not random.

According to the cost of a wrong answer—cost in terms of money, time lost, prestige, and so forth—we may be willing to accept a 1% chance of error, a 5% chance of error, a 10% chance of error, or perhaps even more, in return for knowledge gained about the population. For instance, with an experiment designed to explore an unknown area, any information at all is better than none and we may be willing to risk a larger chance of error in exchange for information. However, in the case of testing drugs, an error could lead to death of people who take that drug as a cure for some disease; consequently, we want the chance of error to be very small.

To decide whether a sequence of runs is in random order or not, we must specify the probability of error that we are willing to accept. It is possible, by methods beyond the scope of this text, to calculate the probability of getting $u$ runs in a group of $N_1$ things of one type and $N_2$ things of another type. For example, if we find 10 runs in a sequence and know that the probability of getting 10 or more runs is quite large if in fact the sequence is random, then we may conclude that the sequence is random based on our sample evidence. For our purposes, we will arbitrarily specify 10% as our level of acceptable error.

Table II in the appendix lists the lower and upper cut-off points for the number of runs among $N_1$ objects of one type and $N_2$ objects of a second type. An observed number of runs that is less than or equal to the first number or greater than or equal to the second number would occur less than ten per cent of the time in a random sequence. In other standard books of statistical tables, the experimenter may find tables similar to Table II which give cut-off points assuming 5% or 1% rather than 10% as the level of acceptable error.

EXAMPLE 6–1. Suppose that 15 missile firings resulted in the following sequence: S FFF S FF SSSS F SSS (the runs in this sequence are underlined). Then $N_1 =$ the number of successes $= 9$, $N_2 =$ the number of failures $= 6$, and $u =$ number of runs $= 7$. From Table II, the cut-off points are 4 and 11. Hence we may conclude that the time sequence of successes and failures is random. The missile is as yet unpredictable, or not reliable for use in a weapon system.

EXAMPLE 6–2. Suppose that a line of children by the drinking fountain has boys and girls in the following order: BBBB GGG BB GGGGG (the runs are underlined). Then $N_1 = 6$, $N_2 = 8$, and $u = 4$. From Table II, the appropriate cut-off points are 4 and 11, so that either we have observed an unusual case or else the order is nonrandom. If there had been two more girls at the end of the line then $N_1 = 6$, $N_2 = 10$, and $u = 4$. From Table II the appropriate cut-off points are 5 and 11, and we would conclude again that the sequence is nonrandom.

**Exercise 6–1:** Suppose that several TV sets are tested for defective parts as they come off an assembly line. Would the following sequence be considered random or nonrandom?

ok, ok, ok, def, ok, ok, ok, ok, def, def, def, def,

**Exercise 6–2:** Suppose that a grower suspects that there is a disease spreading among his tomato vines. He observes the tomato vines in one row and makes note of the diseased (D) and healthy (H) ones in sequence as follows:

D H D D D D D H H H H H

Does he have reason to suspect a contagious disease?

**Exercise 6–3:** A teacher suspects that a certain skill can be learned by observation. She demonstrates the task once and then has each pupil attempt the task while the other pupils watch. She records the successes and failures as follows:

F F F F S F F F S S S S S S S S F

May she conclude that observation helps the children to learn the task?

**Exercise 6–4:** A rat is given a mild electric shock if he turns left at a certain point in a maze and is given a reward of a food pellet

# A ONE-SAMPLE TEST OF RANDOMNESS

if he turns right. Does the following sequence of trials indicate that he has "learned" to turn right?

L L L L R L L R R R R R R R R

It was pointed out earlier in this chapter that if an observed sequence appeared to be nonrandom, it may simply be the case that by chance an unusual sequence had been observed. In the next two exercises, we shall perform experiments which should yield random sequences. However, if you compare your results with those of several of your classmates, you should find that in about 10% of the cases, the conclusion based on the experimental evidence would be that a nonrandom sequence was observed. This result is expected when we allow a 10% chance of error.

**Exercise 6–5:** Toss a coin 10 times and record the sequence of heads and tails. Does your sequence appear to be random or nonrandom?

___  ___  ___  ___  ___  ___  ___  ___  ___  ___     _____

Sequence of tosses                                    Answer

**Exercise 6–6:** Toss a die 10 times and record the sequence of 1's and non-1's. Does your sequence appear to be random or nonrandom?

___  ___  ___  ___  ___  ___  ___  ___  ___  ___     _____

Sequence of tosses                                    Answer

### TERMS TO REMEMBER

**Sequence:** A set of items for which the order of appearance of each item in time or space is important.

**Run:** A sequence of objects of one type that is preceded and followed either by no object at all or by an object of another type.

**Clumping:** A tendency toward long runs.

*chapter 7*

# THE UNIFORM DISTRIBUTION

The researcher performs experiments for the purpose of learning about the basic properties of the population he has sampled. He might fertilize a field of corn with different amounts of fertilizer in order to observe the effects of the different levels upon the yield of corn. He might observe a rat perform in a maze to find whether some manner of experimental treatment has affected the rat's skill. Federal officials might draw a capsule from a large urn in a draft lottery to ascertain the lottery number of a certain birthdate. A gambler might toss a coin or a die several times to decide whether it is fair. For any given experiment, definition of the sample space specifies all possible outcomes or events that could occur.

A single one of these possible outcomes (a single "point" or measurement in the sample space) is called a **simple event.** For example, getting a "2" when a die is tossed once is a simple event. The term "event" is used in general to denote either a simple event or a **compound event,** which is an event that contains two or more sample points. Getting an even number on the toss of a single die would be a compound event because it would involve getting either a "2", a "4", or a "6", and thus any one of three points of the sample space would satisfy the definition of this event. Thus an event is any collection of sample points from the sample space. An event is said to have occurred if the experiment produced the observation of any one of its possible sample points.

## THE UNIFORM DISTRIBUTION

As we have seen before, the model specifies the theoretical distribution of the expected frequencies of all possible outcomes of an experiment. On the other hand, the actual frequency of occurrence of each event that has been observed in an experiment is called the **empirical** (or sampling) **distribution.** We call a coin or die "fair" if it is balanced so that we expect each side to have an equal chance to land upturned. For example, if we say that we will toss a fair coin ten times, then by the terminology "fair coin" we imply that heads and tails are equally likely on any given toss. Thus the mathematical model of a fair coin specifies the chance of "tails" to be 1/2 and the chance, or probability, of "heads" to be 1/2. Therefore, we expect 5 heads and 5 tails. Nevertheless, when we actually toss the coin, we may get five heads and five tails, or seven heads and three tails, or any one of several other possible outcomes.

Thus the sampling distribution, or empirical distribution, that we observe may or may not be the same as that implied by the mathematical model. In fact, we would be quite surprised if we tossed a fair die six times and observed exactly one of each of the six possible values, even though that is the distribution specified by the model. Instead, we are much more apt to get two or maybe three observations of one number and no observations of one or more of the other numbers. If the sampling distribution differs from the model due to chance fluctuations such as these, we call the difference **sampling error.** These sampling errors are not caused by any fault on the part of the experimenter—they merely occur by chance.

EXAMPLE 7–1. If a fair die is tossed one time, then the chance that a "1" appears on the upturned face is exactly one out of six, or 1/6. Similarly, the chance that a "2" appears is 1/6. In fact, we may make a simple probability table like Table 7–1, listing all possible things that could happen (all possible events) and the corresponding chances (probabilities) of their occurrences. We can, if we wish, think of these individual probabilities as the expected frequencies of occurrence of each event whenever a die is tossed one time.

If there are several simple events possible and each is equally likely, then their probabilities are said to have a **uniform distribution,** and they are **uniformly distributed.** A graph illustrating the uniform distribution for this example is shown in Figure 7–1. If we wished, we could label the vertical axis "Expected Frequency" instead of "Probability."

**TABLE 7–1** *POSSIBLE EVENTS AND THEIR PROBABILITIES WHEN A FAIR DIE IS TOSSED ONCE.*

| Number on the upturned face | 1 | 2 | 3 | 4 | 5 | 6 |
|---|---|---|---|---|---|---|
| Probability of that number appearing | 1/6 | 1/6 | 1/6 | 1/6 | 1/6 | 1/6 |

# THE UNIFORM DISTRIBUTION

**Figure 7-1** Graph of the theoretical distribution of events when a fair die is tossed one time.

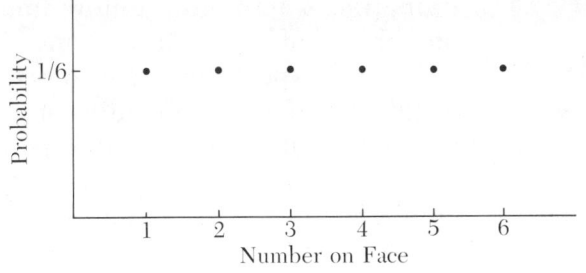

Figure 7-1 is a graphic example of a model, or a theoretical distribution. In general, if n simple events have a uniform distribution, then the probability of each event is $1/n$ and a graph of this distribution would contain n points each located at a height of $1/n$.

EXAMPLE 7-2. The draft lottery, as practiced in the early 1970's, is an example of the uniform distribution. All possible birthdates were listed on slips of paper and placed in capsules. The capsules were placed in a large basket in random sequence; that is, not all December dates at one time, and similarly for the other months. These capsules were then mixed thoroughly. In addition, slips containing the numbers 1 through 366 were placed in random sequence in capsules in a second basket. Two capsules were then drawn, one from each basket. One capsule yielded the birthdate and the other gave its sequence position for the induction of men with that birthdate. After each drawing the capsules were stirred thoroughly before the next set of two capsules was drawn. In this way, the sequence of birthdates drawn was random and the probability of any single birthdate occurring in a particular place in the sequence was 1/366 (February 29 was included). This method of assignment of sequential positions for birthdates is "fair" if birthdates are uniformly distributed. However, birthdates are not uniformly distributed; an obvious exception is February 29.

Many of the examples and exercises that we have done in the last few chapters have illustrated the basic properties of probability. Let us now summarize these basic properties. If A is an event, then the probability of A, sometimes written $P(A)$, has the following properties:

    a. $0 \leq P(A) \leq 1$
    b. $P(A) = 0$ if A is an impossible event
    c. $P(A) = 1$ if A is sure to happen
    d. If the two events A and B have no sample points in common then the probability that either A or B or both will occur is the simple sum of their respective probabilities; that is, $P(A \text{ or } B) = P(A) + P(B)$. As a special case, if A and B are the only two events possible in the sample space, and if A and B have no sample points in common, then $P(A \text{ or } B) = P(A) + P(B) = 1$ so that $P(A) = 1 - P(B)$ and $P(B) = 1 - P(A)$.

## THE UNIFORM DISTRIBUTION

Properties *a* through *c* follow immediately from the definition of the probability of an event in terms of the relative frequency with which we would expect the event to occur. We use property *d*, for example, in calculating the probability of a compound event by merely adding the probabilities of its sample points.

EXAMPLE 7–3. Suppose a bucket has six ping-pong balls in it numbered 1, 2, 3, 4, 5, and 6. If we stir them and draw out one ball, we have performed an experiment. By observing and recording the number on the ball, we specify the outcome of the experiment—a simple event. The model, or theoretical distribution, is given by Table 7–1, and a graph of the model appears as Figure 7–1. Notice that the same model describes both the experiment of this example and the toss of a single fair die. Often the same model applies to several possible physical situations. Now we shall consider some events from this sample space and observe examples of the properties of their probabilities as were summarized above.

A = {drawing a ball numbered "5"}, a simple event; $P(A) = 1/6$.
B = {drawing a ball with an odd number}, a compound event; $P(B) = 3/6$.
C = {drawing a ball numbered "7"}, an impossible event; $P(C) = 0$.
D = {drawing a ball with a number between "1" and "6" inclusive}, a sure thing; $P(D) = 1$.
E = {drawing a ball numbered "6"}, a simple event; $P(E) = 1/6$.
F = {drawing a ball with either a "6" or a "5"}, a compound event; $P(F) = 2/6$.

For all these events (and for any others you can name) $0 \leq P(\text{event}) \leq 1$ (Property *a*). Furthermore, $P(C) = 0$ (Property *b*) and $P(D) = 1$ (Property *c*). $P(F) = P(E \text{ or } A) = P(E) + P(A)$ since A and E have no points in common (Property *d*).

***Exercise 7–1:*** Suppose that five people (call them A, B, C, D, E) are drawing straws to see who pays for the Cokes. If there is one short straw and four long straws, what are the chances that A will have to pay for the Cokes? _____ What are the chances B will have to pay? _____ C? _____ D? _____ E? _____. Is this an example of the uniform distribution? _____ Make a

## THE UNIFORM DISTRIBUTION

probability table similar to Table 7-1 listing the sample space and the probability of each event of the sample space.

Sketch a figure similar to Figure 7-1 illustrating the probability distribution.

**Exercise 7-2:** Suppose you toss a fair coin once. What are the chances that it lands heads? _____ What are the chances that it lands tails? _____ Is this an example of a uniform distribution? _____ Make a probability table similar to Table 7-1 listing the sample space and the probability of each event.

# THE UNIFORM DISTRIBUTION

Sketch a figure similar to Figure 7-1 illustrating the probability distribution.

**Exercise 7-3:** (a) Toss a die 6 times and record the number of times each number occurs in the following table:

| Number | Frequency of Occurrence |
|--------|------------------------|
| 1 | |
| 2 | |
| 3 | |
| 4 | |
| 5 | |
| 6 | |

(b) Now graph the empirical distribution:

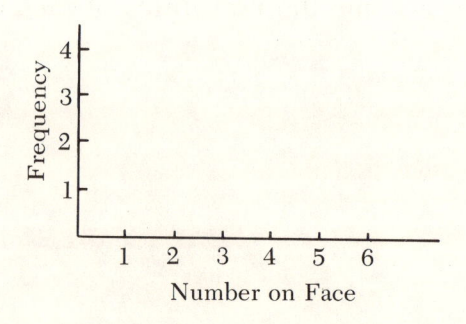

**THE UNIFORM DISTRIBUTION**

(c) Does this graph resemble Figure 7–1? _____ Why or why not? _____

_____

(Notice that we could have labeled the vertical axis with the relative frequencies of 1/6, 2/6, 3/6, etc. These would have described the same properties in the empirical distribution as are described by the model's probabilities.)

**Exercise 7–4:** (a) Go to Table II in the appendix and choose any starting point. Take 6 observations in the following manner: If the first digit observed is either a "1", a "2", a "3", a "4", a "5", or a "6", write it down, but if it is a "0", a "7", an "8", or a "9", ignore it and go to the next digit. Record the number of times each digit occurs in the following table:

| Number | Frequency of Occurrence |
|--------|------------------------|
| 1      |                        |
| 2      |                        |
| 3      |                        |
| 4      |                        |
| 5      |                        |
| 6      |                        |

(b) Now graph the empirical distribution:

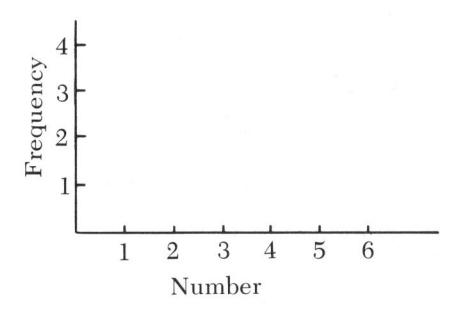

(c) Does the graph resemble Figure 7–1? _____ Why or why not? _____

_____

The empirical distribution, the sample distribution of 1's, 2's,

## THE UNIFORM DISTRIBUTION

3's, 4's, 5's, and 6's actually observed, may differ from the model, or expected, distribution. The chances of getting a "2" are 1 out of 6, but when a die is tossed 6 times (or when we observe a digit from the random number table) the experimenter could get anywhere from zero to six "2's"; in other words, there could be sampling error.

**Exercise 7-5:** (a) Put 15 dice in a cup or jar, or hold them in your hand, and toss them all at once. In the following table, record the number of times each number occurs by listing work marks or tally marks which indicate the counts for each number showing on the dice. Repeat the toss four times for a total of 60 observations. After the experiment is completed, record the total frequencies indicated by the work marks.

| Number | Work Marks | Total Frequencies |
|--------|------------|-------------------|
| 1 | | |
| 2 | | |
| 3 | | |
| 4 | | |
| 5 | | |
| 6 | | |

(b) Now graph the empirical distribution on the following graph:

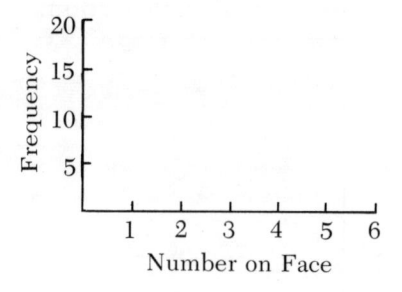

(c) Does this graph resemble Figure 7-1 more closely than did those of Exercises 7-3 or 7-4? _____ Why or why not? _____

_____

**Exercise 7-6:** (a) Put 30 dice in a cup or jar and toss them all at

THE UNIFORM DISTRIBUTION

once. Record the observations as outlined in Exercise 7–5, and repeat once for a total of 60 observations.

| Number | Work Marks | Total Frequencies |
|--------|------------|-------------------|
| 1 | | |
| 2 | | |
| 3 | | |
| 4 | | |
| 5 | | |
| 6 | | |

(b) Now graph the empirical distribution on the following graph:

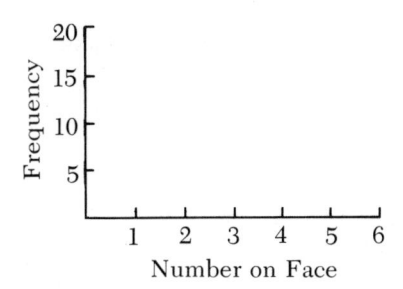

(c) Does this graph resemble that of Figure 7–1? _____

Does it resemble those of Exercises 7–3 or 7–4? _____

Why or why not? _____

_____

Because random numbers are uniformly distributed, taking observations from a random numbers table (as indicated in Exercise 7–4) is equivalent to tossing a fair die (as in Exercise 7–3). Also, if the dice are all fair, the experimenter is performing equivalent experiments if he tosses one die 60 times, or four dice 15 times, or fifteen dice 4 times, or thirty dice 2 times, since all of these experimental treatments conform to the same model. Thus all should produce similar empirical distributions, even though the sampling errors that are always present will cause individual variations.

Now try this "brain teaser:" suppose you have one fair die, which has the numbers 1, 2, 3, 4, 5, and 6 on its sides; and you also have one

## THE UNIFORM DISTRIBUTION

blank fair die on which you may put any numbers. What numbers should you put on the six sides of the blank die in order that the numbers 1, 2, 3, 4, 5, 6, 7, 8, 9, 10, 11, or 12 each have an equal chance of occurring as the total number of dots appearing on the upturned faces of the dice? (Hint: sides may be blank or may have some numbers repeated.)

### TERMS TO REMEMBER

**Simple Event:** An event or subset of the sample space which contains only a single measurement or item from the sample space.

**Compound Event:** An event which contains two or more measurements or items from the sample space.

**Empirical Distribution:** A listing of the actual observed frequencies (or relative frequencies, or percentage frequencies) of occurrence of all possible events.

**Sampling Error:** The deviation of a sample result from its expected outcome due to chance fluctuations.

**Uniform Distribution:** The model in which all sample events are equally likely, that is, have the same probability.

chapter 8

# THE BERNOULLI DISTRIBUTION

When a fair coin is tossed once, either "heads" will appear or "tails" will appear. Each of these two possible events is equally likely, so that the graph of the theoretical distribution is illustrated by Figure 8–1; this is a uniform distribution.

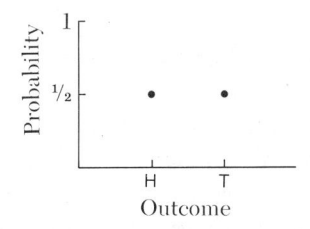

**Figure 8–1**  Graph of the distribution of a fair coin.

Suppose now that the coin is not fair but has only a 1/6 chance of landing "heads" and hence a 5/6 chance of being "tails." Then the mathematical model would be graphed as in Figure 8–2.

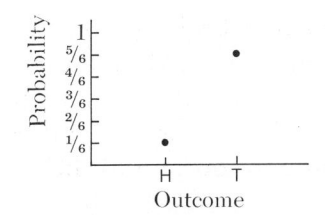

**Figure 8–2**  Graph of the distribution of an unfair coin.

# THE BERNOULLI DISTRIBUTION

Each time we toss a coin one of two possible events may occur: a head or a tail. If we get a head on one toss then we cannot also get a tail on the same toss, and vice versa. If on any one performance of the experiment the occurrence of one event precludes the occurrence of the other, then we call the events **mutually exclusive** or **disjoint.**

Tossing either a fair or an unfair coin is an example of a Bernoulli experiment. In general if an experiment has only two possible outcomes and these outcomes are mutually exclusive, then the experiment is called a Bernoulli experiment or **Bernoulli trial,** and the distribution of probabilities of an experiment of this type is called a **Bernoulli distribution.** The Bernoulli distribution is also *uniform* if the two possible outcomes are equally likely.

Some examples of Bernoulli trials which occur in our everyday experiences are: getting a hit or not getting a hit when a player comes to bat, the girl answering "yes" or "no" when asked for a date, pass or fail on a test. Notice that in each case the two events are mutually exclusive. If one has occurred on a particular trial, then the other cannot occur on that same trial.

There are also many examples of events that are not mutually exclusive. A team may either lose or fail to lose a football game; when they fail to lose, they could either win or tie. If either an odd number or a number less than 3 appears on the toss of a single die, we recognize that both events contain the item "1" as a possibility. If we hope to catch a bass or a fish on a fishing trip, we have specified two possible events that are not disjoint, because a bass is a fish.

**Exercise 8-1:** Which of the following describe two mutually exclusive events?
 a. Getting an even number or getting a number less than 3 when a die is tossed once.
 b. Getting a passing grade or getting an A on a theme.
 c. Getting a total of "5" on two dice or getting a "1" on one of the dice when two dice are tossed together.
 d. Finding a robin or finding a bird on a nature study tour.

Answer _____

**Exercise 8-2:** Of the following, which pairs of events are mutually exclusive?
 a. Finding a bluebird or finding a bird on a nature study tour.
 b. Getting an A or getting an E on a theme.
 c. Tossing a "3" or tossing a "5" on a single die.
 d. Tossing a total of "7" with a pair of dice or tossing two dice and having a "4" come up on one of the dice.

Answer _____

**Exercise 8-3:** Give 4 more examples of Bernoulli trials.

**Exercise 8-4:** Describe 4 situations in which Bernoulli trials arise.

**Exercise 8-5:** Toss a coin once and indicate by a graph the frequency observed for "head" and "tail."

**Exercise 8-6:** Toss a die once and graph the frequencies of the "1" and "not 1" outcomes ("not 1" would occur whenever either a "2," a "3," a "4," a "5," or a "6" is observed).

**Exercise 8-7:** Toss a coin 10 times and graph the frequencies of H and T.

## THE BERNOULLI DISTRIBUTION

**Exercise 8-8:** Toss a die 10 times and graph the frequencies of "1" and "not-1" outcomes.

**Exercise 8-9:** Toss a coin 100 times and graph the frequency with which you observe "heads" and "tails".

**Exercise 8-10:** Toss a die 100 times and graph the frequencies of "1" and "not-1" outcomes.

**Exercise 8-11:** Even though the vertical scale of Figure 8-1 is in terms of a probability whereas those of the graphs in the exercises call for the frequency of occurrence, still the model for the experiments in Exercises 8-5, 8-7, and 8-9 is given by that illustrated in Figure 8-1. Which of the graphs in Exercises 8-5, 8-7, and 8-9 appear most like Figure 8-1? _____ Why? _____

**Exercise 8-12:** Even though the vertical scale of Figure 8-2 is in terms of a probability whereas those of the graphs in the exercises

## THE BERNOULLI DISTRIBUTION

call for the frequency of occurrence, still the model for the experiments in Exercises 8-6, 8-8, and 8-10 is given by that illustrated in Figure 8-2. Which of the graphs in Exercises 8-6, 8-8, and 8-10 appear most like Figure 8-2? _____ Why? _____

_____

***Exercise 8-13:*** Toss a coin 20 times in such a way that it is no longer a "fair" toss. Graph the results.

How did you toss the coin "unfairly?" _____

_____

***Exercise 8-14:*** Toss a die 30 times in such a way that it is no longer a "fair" toss. Graph the results.

How did you toss the die "unfairly?" _____

_____

Sometimes an experimenter gathers and records data that are either larger or smaller on the average than they should be according to the model involved. These measurements are said to be biased. Exercises 8-13 and 8-14 are examples of experimental bias.

**Bias** is a systematic deviation of experimental results from the outcome expected according to the model specified—a deviation caused by something other than the experimental treatment. Bias can occur from some physical cause, such as an improperly calibrated or faulty measuring instrument. It can occur also because the experi-

# THE BERNOULLI DISTRIBUTION

menter is so confident he knows the outcome of an experiment before it is performed that he "observes" erroneous measurements. A faulty observational technique can also cause bias.

As an example of a physical cause for bias, suppose that waist measurements of seventh grade children are being taken by means of a tape measure; and, because of repeated use, the first one-half inch of the tape tears off. If this is not noticed and the experimenter continues measuring, all subsequent measurements will have a consistent error or bias of one-half inch. Other examples of bias due to a systematic deviation in a measuring instrument would be an improperly balanced scale for measuring weights, or a teacher's examination "key" which has one or more incorrect answers.

As an example of bias due to preconceived notions, suppose a nurse is convinced that a patient will respond better to a Drug A than to Drug B. If the nurse who administers the drugs is the same one who observes and records the response, then regardless of the true effect of the drugs, she might easily "see" what she expects to see rather than take a careful objective visual measurement. Errors caused by faulty judgment on the part of the experimenter are far more common when response is measured subjectively (like a nurse observing a patient's response) rather than objectively (like recording the weight of an object after reading the value on a balanced scale).

An example of a faulty observational technique could be reading a bathroom scale or a speedometer from a position to one side or the other rather than from a position directly perpendicular to the dial. This would result in parallax, which is the difference in apparent position of the needle or dial when seen from different positions.

Because it is sometimes difficult to be completely objective, a **blind experiment** is often designed. This is an experiment in which the person who performs the experiment does not know which treatments or physical manipulations have been applied to the subjects or pieces of experimental material. For example, in blind experiments, the physician or nurse who observes the results of various drugs on different patients does not know what drug was administered to each patient, or the technician who measures the yield of corn from experimental plots did not apply the different levels of fertilizer to the plots and thus does not know which plot received a particular level of fertilizer. Blind experiments promote objective, unbiased observations.

### TERMS TO REMEMBER

**Mutually Exclusive Events:** Events which have the property that the occurrence of one implies that the other cannot occur at the same time. Events which have no common sample points.

**Disjoint Events:** Mutually exclusive events.

**Bernoulli Trial:** An experiment which has only two possible mutually exclusive outcomes.

**Bernoulli Distribution:** The theoretical model for a Bernoulli trial.

**Bias:** A systematic deviation of experimental results from the outcome expected under the model, a deviation which is caused by factors other than treatment effects.

**Blind Experiment:** An experiment for which the person who records the experimental results does not know what treatments were applied to the experimental material.

chapter 9

# THE BINOMIAL DISTRIBUTION

In the last chapter we discussed the experiment of tossing a fair coin once. Consider now the experiment of tossing a fair coin several times in succession. By the "outcome" of this experiment, we shall mean simply the total number of heads observed in all of the tosses.

When we are concerned with two disjoint events we often call them "success" and "failure," or merely S and F, where **success** means an outcome of one specified type, and **failure** means an outcome of a second type. By calling "heads" the "success," we do not imply that "tails" are less desirable even though "tails" would be labeled "failure." If the outcome of one event or trial does not affect that of any preceding or succeeding event or trial, then the events or trials are called **independent.**

For example, suppose a fair coin is tossed three times. What are all possible outcomes? We may observe no heads (TTT), one head (HTT, THT, or TTH), two heads (HHT, HTH, or THH), or three heads (HHH). These outcomes could be listed as shown in Table 9–1. Since the coin is fair and since the outcomes of one toss do not affect those of any other toss, we may assume that the eight possible elementary events from the sample space are equally likely. Thus the probability of any given outcome is merely the ratio of the number of ways it can occur divided by eight, the total number of possible outcomes. If we consider the eight possible ordered sequences as listed in Table 9–1, they have a uniform distribution. However, we agreed

# THE BINOMIAL DISTRIBUTION

**TABLE 9-1** POSSIBLE OUTCOMES (AND THEIR PROBABILITIES) WHEN A COIN IS TOSSED THREE TIMES.

| Outcomes | Ordered Sequences | Number of Elementary Events | Probability of Outcomes |
|---|---|---|---|
| 0 Head | TTT | 1 | 1/8 |
| 1 Head | HTT or THT or TTH | 3 | 3/8 |
| 2 Heads | HHT or HTH or THH | 3 | 3/8 |
| 3 Heads | HHH | 1 | 1/8 |

to define as our desired outcome the total number of heads without regard to the order in which they occurred, and the distribution of the probabilities of the events 0, 1, 2, or 3 heads will not be uniform.

If we toss a coin ten times, we may observe no heads, or one head, or any number of heads up to and including ten heads. The distribution of the probabilities of outcomes in experiments such as these is called the **binomial distribution.** The binomial distribution is the distribution of $n$ independent Bernoulli trials, each of which has the same probability of success. We know from Chapter 8 that a Bernoulli trial is one for which two possible events can occur. We may specify one as the "success" and the other as the "failure".

The probabilities of the individual outcomes for the binomial distribution may be obtained in three different ways: (1) Directly, by listing and counting the possibilities, as was done in Table 9-1, (2) using the Pascal Triangle (see Table 9-2), or (3) using the formidable formula for the binomial expansion which is frighteningly familiar to the high school algebra class. These methods are equivalent, and of the three methods, the Pascal Triangle is often the simplest.

Consider the Pascal Triangle in the third column of Table 9-2. The "outside" numbers are all "ones." The second number in any row is the row number. The sum of the numbers in any row is 2 to the power of the row number. For row 2, there were 2 tosses, its row in the triangle is "1 2 1", and $1+2+1=4=2^2$. Notice the symmetry of

**TABLE 9-2** THE PASCAL TRIANGLE.

| Row | Number of Tosses | Pascal Triangle | Total of the Numbers in the Rows |
|---|---|---|---|
| 1 | 1 | 1  1 | $2 = 2^1$ |
| 2 | 2 | 1  2  1 | $4 = 2^2$ |
| 3 | 3 | 1  3  3  1 | $8 = 2^3$ |
| 4 | 4 | 1  4  6  4  1 | $16 = 2^4$ |
| 5 | 5 | 1  5  10  10  5  1 | $32 = 2^5$ |

# THE BINOMIAL DISTRIBUTION

the numbers in each row: for instance, in the fifth row, a "1" is the first and last, a "5" is second and second from last, and so on. Finally, notice that any "inside" number is the sum of the two numbers directly above it: in row 5 the second number—a "5"—is the sum of the "1" and "4" which appear directly above it, and the third number—a "10"—equals $4+6$. The first row of the Pascal Triangle indicates that if a coin is tossed once, no heads may be obtained only 1 way and one head only 1 way. For this single toss there are $1+1=2$ possible outcomes. The second row indicates that when a coin is tossed twice, no heads may be obtained 1 way (TT), one head in 2 ways (HT or TH), and two heads in 1 way (HH), yielding $1+2+1=4=2^2$ possible outcomes.

Notice the similarity of the third row to Table 9-1. From the third row we see that there are 8 possible outcomes $(1+3+3+1)$ when a coin is tossed 3 times: no heads in 1 way, one head in 3 ways, two heads in 3 ways, and three heads in 1 way. Thus the chance of getting no heads, or all tails, when a fair coin is tossed three times is 1 out of 8 or 1/8. The chance of getting one head in three tosses is 3 out of 8 or 3/8.

EXAMPLE 9-1. Read the answers to the following questions from the Pascal Triangle in Table 9-2 (answers are in parentheses):
1. How many outcomes can occur when a coin is tossed 4 times? ($2^4 = 16$)
2. How many outcomes can occur when a coin is tossed twice? ($2^2 = 4$)
3. If a coin is tossed 3 times, in how many ways may you get 2 heads? (3) 3 heads? (1)
4. If a coin is tossed 5 times, in how many ways can you get at least 3 heads? ($10+5+1=16$) At most 2 heads? ($1+5+10=16$) At most 1 head? ($1+5=6$) No heads? (1)
5. What are the chances of getting 2 heads in 4 tosses of a fair coin? (6/16) Of getting 1 head in 2 tosses? (1/2)
6. What are the chances of having 4 boys in a family of 5 children? (5/32)
7. What is the probability of having 4 corn seeds all sprout if for each seed the chance of sprouting is 0.5? (1/16)
8. What are the chances of having at least two girls in a family with 4 children? (11/16)

Notice that for the last three questions in Example 9-1 we were not considering coin tossing, but we were able to use the coin tossing model outlined in this chapter. This model has wide applicability to several practical types of experimental situations; the only requirement is that the chance of success on any one trial be 1/2.

**THE BINOMIAL DISTRIBUTION**

*Exercise 9-1:* Write rows 6, 7, and 8 of the Pascal Triangle.

From row 6, how many ways may we get exactly 2 heads when a coin is tossed 6 times? _____ From row 8, in how many ways may we get 6 or more heads when a coin is tossed 8 times? _____

*Exercise 9-2:* Write rows 6, 7, and 8 of the Pascal Triangle.

From row 6, how many ways may we get exactly 4 heads when a coin is tossed 6 times? _____ In how many ways may we get 2 or fewer heads when a coin is tossed 8 times? _____

*Exercise 9-3:* If a coin is tossed 4 times, what are the chances of getting 2 heads? _____ If it is tossed 5 times, what are the chances of getting 3 or more heads? _____ How many possible outcomes are there when a coin is tossed 5 times? _____

*Exercise 9-4:* If a coin is tossed 6 times, what are the chances of getting 3 heads? _____ If it is tossed 7 times, what are the chances of getting 4 or more heads? _____ How many possible outcomes are there when a coin is tossed 7 times? _____

In addition to the Pascal Triangle, we may use binomial expan-

## THE BINOMIAL DISTRIBUTION

sions such as $(T + H)$, $(T + H)^2$, $(T + H)^3$, and in general, $(T + H)^n$ to calculate the number of ways in which each outcome of a binomial experiment can occur. From these we can also calculate the probability of the outcome. Computing the binomial expansion for high powers such as $(T + H)^{10}$ can be quite time consuming. Hence, it is often easier to use the Pascal Triangle to calculate probabilities for a binomial experiment.

The expression $(T + H)^n$, which may be more familiar in the form $(x + y)^n$ or $(a + b)^n$, is simply equal to the product of $n$ factors $(T + H)$. For example,

$$(T + H)^1 = (T + H) = T + H$$

$$(T + H)^2 = (T + H)(T + H) = T^2 + 2TH + H^2$$

$$(T + H)^3 = (T + H)(T + H)(T + H) = (T^2 + 2TH + H^2)(T + H)$$
$$= T^3 + 3T^2H + 3TH^2 + H^3$$

and

$$(T + H)^n = \underbrace{(T + H)(T + H) \ldots (T + H)}_{n \text{ factors}} = (T + H)^{n-1}(T + H)$$

There is a complicated formula that can be used to calculate directly the expansion of any binomial power; however, we will not consider this formula here. We will obtain the expansions of the binomial powers simply by carrying out the necessary multiplications, as we did above.

As we verified for the cases $n = 1$, $n = 2$, and $n = 3$, the binomial expansion of $(T + H)^n$ consists of a sum of $n + 1$ terms, each of which is a product of a whole number with some T's and H's. The whole number (called the coefficient) in each term is equal to the number of ways in which one can obtain, in $n$ tosses of a coin, as many T's (tails) and H's (heads) as there are in that term. Thus, from the expansion of $(T + H)^3$, we see that in three tosses of a coin, there is one way to

**TABLE 9-3** BINOMIAL EXPANSION FOR $n = 1, 2,$ AND $3$.

| No. of Tosses | Expansion | No. of Events | Listing of Events | Pascal Triangle |
|---|---|---|---|---|
| 1 | $(T + H)^1 = T + H$ | 2 | {T} or {H} | 1 1 |
| 2 | $(T + H)^2 = T^2 + 2TH + H^2$ | 4 | {TT}; {TH, HT}; {HH} | 1 2 1 |
| 3 | $(T + H)^3 = T^3 + 3T^2H + 3TH^2 + H^3$ | 8 | {TTT}; {TTH, THT, HTT}; {THH, HTH, HHT}; {HHH} | 1 3 3 1 |

## THE BINOMIAL DISTRIBUTION

**TABLE 9-4** COMPARISON OF THE METHODS OF THE PASCAL TRIANGLE AND THE BINOMIAL FORMULA TO OBTAIN PROBABILITIES IN A BINOMIAL EXPERIMENT.

| No. of Tosses | Pascal Triangle | Probabilities (Assuming Fair Coins) | Binomial Formula | Probabilities (Assuming Fair Coins) |
|---|---|---|---|---|
| 1 | 1 1 | 1/2  1/2 | $T + H$ | 1/2  1/2 |
| 2 | 1 2 1 | 1/4  1/2  1/4 | $T^2 + 2TH + H^2$ | 1/4  1/2  1/4 |
| 3 | 1 3 3 1 | 1/8  3/8  3/8  1/8 | $T^3 + 3T^2H + 3TH^2 + H^3$ | 1/8  3/8  3/8  1/8 |

obtain three tails and no heads ($T^3$), three ways to obtain two tails and a head ($T^2H$), three ways to obtain a tail and two heads ($TH^2$), and one way to obtain no tails and three heads ($H^3$). Notice that these are the same results we obtained by using the Pascal Triangle. These results are summarized in Table 9-3 in a manner which illustrates the connection between this method and that of the Pascal Triangle.

One of the basic uses of both the Pascal Triangle and the binomial formula is to find the probability of each possible individual outcome or event that can occur in observing the results of the $n$ independent Bernoulli trials. For example, by dividing each entry of a row of the Pascal Triangle by the sum of the elements in its row, we have a listing of the probabilities for each event, as is shown in Table 9-4. Similarly, if we substitute for H and T in the binomial formula the probability of obtaining heads or tails (which for a fair coin is 1/2), we can calculate the probability of each individual event, as is done in the last column of Table 9-4. Of these two methods, the latter is more general, since if we should toss an unfair coin—say one for which $P(H) = 1/4$ and $P(T) = 3/4$—we can still use this method to find the probability of each individual outcome as is demonstrated in Table 9-5. On the other hand, if we simply divide each element in the Pascal Triangle by the total of the entries for the row, we assume that we are considering only fair coins.

**TABLE 9-5** USE OF THE BINOMIAL FORMULA TO OBTAIN PROBABILITIES IN A BINOMIAL EXPERIMENT IN WHICH THE PROBABILITIES FOR SUCCESS AND FAILURE ARE NOT EQUAL: $P(H) = 1/4$ AND $P(T) = 3/4$.

| No. of Tosses | Binomial Formula | Substitution | Probabilities |
|---|---|---|---|
| 1 | $T + H$ | $\frac{3}{4} + \frac{1}{4}$ | $\frac{3}{4}$  $\frac{1}{4}$ |
| 2 | $T^2 + 2TH + H^2$ | $\left(\frac{3}{4}\right)^2 + 2\left(\frac{3}{4}\right)\left(\frac{1}{4}\right) + \left(\frac{1}{4}\right)^2$ | $\frac{9}{16}$  $\frac{6}{16}$  $\frac{1}{16}$ |
| 3 | $T^3 + 3T^2H + 3TH^2 + H^3$ | $\left(\frac{3}{4}\right)^3 + 3\left(\frac{3}{4}\right)^2\left(\frac{1}{4}\right) + 3\left(\frac{3}{4}\right)\left(\frac{1}{4}\right)^2 + \left(\frac{1}{4}\right)^3$ | $\frac{27}{64}$  $\frac{27}{64}$  $\frac{9}{64}$  $\frac{1}{64}$ |

## THE BINOMIAL DISTRIBUTION

**Exercise 9-5:** By multiplying $T^3 + 3T^2H + 3TH^2 + H^3$ by $(T+H)$, expand $(T+H)^4$. List all sixteen events that could occur when a coin is tossed 4 times and use the list to check whether you obtained the correct expansion. Do the coefficients of your expansion agree with those in the fourth row of the Pascal Triangle? _____

**Exercise 9-6:** By multiplying $T^4 + 4T^3H + 6T^2H^2 + 4TH^3 + H^4$ by $(T+H)$, expand $(T+H)^5$. List all thirty-two events that could occur when a coin is tossed 5 times and use the list to check whether you obtained the correct expansion. Do the coefficients of your expansion agree with those in the fifth row of the Pascal Triangle? _____

# THE BINOMIAL DISTRIBUTION

**Exercise 9-7:** To find the probability of all possible events when a fair coin is tossed four times, replace the "H" and "T" in the expansion of Exercise 9-5 by their probabilities, namely 1/2 and 1/2. Fill in the following table with your results:

| Event | no heads | 1 head | 2 heads | 3 heads | 4 heads |
|---|---|---|---|---|---|
| Probability | | | | | |

**Exercise 9-8:** To find the probability of all possible events when a fair coin is tossed five times, replace the "H" and "T" in the expansion of Exercise 9-6 by their probabilities, namely 1/2 and 1/2. Fill in the following table with your results:

| Event | no heads | 1 head | 2 heads | 3 heads | 4 heads | 5 heads |
|---|---|---|---|---|---|---|
| Probability | | | | | | |

**Exercise 9-9:** Toss 3 coins and observe the results: if you get TTT, put an X in the "3 tails" cell for row "1"; if you get THT or HTT or TTH, put an X in the "2 tails" cell for row "1"; and so forth. Repeat the experiment 23 times and fill in the results in Table 9-6.

*TABLE 9-6* FREQUENCY TABLE FOR EXERCISE 9-9.

| Experiment Number | 0 TAIL | 1 TAIL | 2 TAILS | 3 TAILS |
|---|---|---|---|---|
| 1 | | | | |
| 2 | | | | |
| 3 | | | | |
| 4 | | | | |
| 5 | | | | |
| 6 | | | | |
| 7 | | | | |
| 8 | | | | |
| 9 | | | | |

## THE BINOMIAL DISTRIBUTION

**TABLE 9-6** FREQUENCY TABLE FOR EXERCISE 9-9 (Continued).

| Experiment Number | Outcomes | | | |
|---|---|---|---|---|
| | 0 TAIL | 1 TAIL | 2 TAILS | 3 TAILS |
| 10 | | | | |
| 11 | | | | |
| 12 | | | | |
| 13 | | | | |
| 14 | | | | |
| 15 | | | | |
| 16 | | | | |
| 17 | | | | |
| 18 | | | | |
| 19 | | | | |
| 20 | | | | |
| 21 | | | | |
| 22 | | | | |
| 23 | | | | |
| 24 | | | | |
| TOTALS (Empirical Distribution) | | | | |
| AVERAGE OBSERVED (Empirical Relative Frequencies) | | | | |
| AVERAGE EXPECTED (Probabilities Given by Model) | 3/24 | 9/24 | 9/24 | 3/24 |

For this exercise, the quantity to be filled in in the next-to-last row, the average observed, is equal to the total number of X's in the column divided by 24, and specifies the empirical relative frequencies. From the Pascal Triangle we see that in every set of 8 trials we would expect one TTT, 3 each of the outcomes having exactly 2 H and exactly 2 T,

## THE BINOMIAL DISTRIBUTION

and one outcome of HHH. Thus in the 24 trials, the model specifies three tails 3 times, two tails 9 times, one tail 9 times, and no tails 3 times. Were your empirical relative frequencies the same as the probabilities given by the model? _____ Why or why not?

_____

_____

If you performed the experiment 240 times, how many TTT outcomes would you expect? _____ Would you be surprised if in 240 trials there were 29 TTT outcomes? _____ If there were 31 TTT outcomes? _____ If there were 150 TTT outcomes? _____ Why or why not? _____

_____

Sketch a graph of the average of the outcomes of your 24 experiments.

**Exercise 9–10:** Toss 4 coins and observe the results: if all four land "tails", put an X in the "4 tails" cell of row "1"; if 3 land "tails" and 1 lands "heads", put an X in the "3 tails" cell for row "1"; and so forth. Repeat the experiment 31 times and record the results in Table 9–7.

81

# THE BINOMIAL DISTRIBUTION

**TABLE 9-7** FREQUENCY TABLE FOR EXERCISE 9-10.

| Experiment Number | 4 TAILS | 3 TAILS | Outcomes<br>2 TAILS | 1 TAIL | 0 TAIL |
|---|---|---|---|---|---|
| 1 | | | | | |
| 2 | | | | | |
| 3 | | | | | |
| 4 | | | | | |
| 5 | | | | | |
| 6 | | | | | |
| 7 | | | | | |
| 8 | | | | | |
| 9 | | | | | |
| 10 | | | | | |
| 11 | | | | | |
| 12 | | | | | |
| 13 | | | | | |
| 14 | | | | | |
| 15 | | | | | |
| 16 | | | | | |
| 17 | | | | | |
| 18 | | | | | |
| 19 | | | | | |
| 20 | | | | | |
| 21 | | | | | |
| 22 | | | | | |
| 23 | | | | | |

**TABLE 9-7** FREQUENCY TABLE FOR EXERCISE 9–10 (Continued).

| Experiment Number | 4 TAILS | 3 TAILS | Outcomes<br>2 TAILS | 1 TAIL | 0 TAIL |
|---|---|---|---|---|---|
| 24 | | | | | |
| 25 | | | | | |
| 26 | | | | | |
| 27 | | | | | |
| 28 | | | | | |
| 29 | | | | | |
| 30 | | | | | |
| 31 | | | | | |
| 32 | | | | | |
| TOTALS (Empirical Distribution) | | | | | |
| AVERAGE OBSERVED (Empirical Relative Frequencies) | | | | | |
| AVERAGE EXPECTED (Probabilities Given in Model) | 2/32 | 8/32 | 12/32 | 8/32 | 2/32 |

For this exercise, the quantity to be filled in in the next-to-last row, the average observed, is equal to the total number of X's in the column divided by 32, and specifies the empirical relative frequencies. In every set of 16 trials, we see from the Pascal Triangle that we would expect one TTTT, four outcomes which have exactly 3 tails, six of the outcomes having 2 tails, four having only 1 tail, and one HHHH. Thus in the 32 trials, the model specifies that we should get 4 tails twice, 3 tails eight times, 2 tails twelve times, 1 tail eight times, and no tails twice. Were your empirical relative frequencies the same as the probabilities given by the model? _____ Why or why not?

If you performed the experiment 320 times, how many TTTT outcomes would you expect? _____ Would you be surprised if

# THE BINOMIAL DISTRIBUTION

in 320 trials you observed 19 TTTT outcomes? _____ If you observed 21 TTTT outcomes? _____ If you observed 150 TTTT outcomes? _____ Why or why not? _____

_____

Sketch a graph of the average of the outcomes of your 32 experiments.

**TERMS TO REMEMBER**

**Success:** An outcome of a specified type, such as one of the two outcomes of a Bernoulli trial.

**Failure:** The other outcome of a pair of disjoint events for which the first is called the "success."

**Independent Events:** Given two events A and B, if the occurrence of A does not affect whether or not B will occur, then A and B are said to be independent events.

**Binomial Distribution:** The distribution of probabilities of the total number of successes in $n$ independent Bernoulli trials, each of which has the same probability of success.

chapter 10

# THE BINOMIAL TEST

In Chapter 9, we calculated the chances of observing 0 heads, 1 head, 2 heads, or 3 heads in three tosses of a fair coin to be 1/8 = .125, 3/8 = .375, 3/8 = .375, and 1/8 = .125, respectively. Compare these values with those given for $n =$ the number of successes when $N =$ number of trials $= 3$ and $p = P(\text{success}) = .5$ in Table III of the appendix. We found that we could calculate these probabilities by simplifying the expression obtained when we let $P(H) = 1/2$ and $P(T) = 1/2$ in the binomial expansion $T^3 + 3T^2H + 3TH^2 + H^3$. In other words, they were $(1/2)^3 + 3(1/2)^2(1/2) + 3(1/2)(1/2)^2 + (1/2)^3 = 1/8 + 3/8 + 3/8 + 1/8$. These values have already been calculated for us for several different values of $P(\text{success})$ and are listed in Table III.

Suppose that instead of H or T, each with probability of 1/2, we are concerned with whether a product on an assembly line is defective or non-defective; further suppose that we know that the chance of producing a defective is $1/10 = 0.1$ when the machines are in proper working order. Since the chance of a defective is 0.1, the chance of a non-defective would be 0.9. Because we want a general expression to deal with defectives and non-defectives instead of heads and tails, we shall use the binomial expansion in the form $p^3 + 3p^2q + 3pq^2 + q^3$. With this expansion we may calculate the chance of none, one, two, or three non-defectives in a group of three items on the assembly line as $P(\text{no non-defectives}) = (.1)^3(.9)^0 = .001$; $P(\text{one non-defective}) = 3(.1)^2(.9)^1 = .027$; $P(2 \text{ non-defectives}) = 3(.1)(.9)^2 = .243$;

## THE BINOMIAL TEST

and $P(3 \text{ non-defectives}) = (.9)^3 = .729$. Similarly, the probability of none, one, two, or three defectives would be .729, .243, .027, and .001 respectively. Compare these values with $N = 3$ and $p = .1$ in Table III of the appendix. Another way of expressing these probabilities is to say that if the probability of a defective item is 0.10, then one time in a thousand we would expect all three items to be defective. This occurrence would be unusual, or rare. Twenty-seven times in a thousand we would expect exactly two of the three items to be defective, still an unusual occurrence. We would expect exactly one defective 243 times out of a thousand or 24.3% of the time, and 72.9% of the time we would expect to find no defectives in three items taken off the assembly line. Thus we see that if $p = P(\text{success})$ and $q = 1 - p = P(\text{failure})$ in any type of binomial experiment, we may calculate the probability of any number of successes using the same binomial expansion that we have already studied for the coin tossing model. However, it is much easier merely to read the appropriate values from Table III.

Events which occur, on the average, only once in a thousand trials or even 27 times in a thousand trials, such as we discussed in the preceding paragraph, are appropriately called rare events. A **rare event** is one which is unexpected or out of the ordinary, one that has a small probability of occurring. Usually an event which, according to the model, would happen less than 1% of the time or perhaps less than 5% of the time is considered a rare event. Thus an event might be considered rare if it could be expected to happen only once in 100 trials, or only 5 times in 100 trials. Prior to experimentation the researcher chooses this arbitrary "cutoff point." Once chosen, it remains fixed throughout the experiment.

The experimenter may wish to observe rare events or extreme events or both. An **extreme outcome,** or extreme event, is one which is an outlier relative to the size of the observations expected under the model. In order to visualize the differences and similarities between extreme events and rare events, we shall consider the histogram of the binomial model for $N = 10$ and $p = 0.4$ (see Table III for the appropriate probabilities) as shown in Figure 10–1. Suppose that an experimenter, based on his knowledge of the field and his intentions concerning the use of the experimental results, decides to specify 0.05 (or 5%) as the level indicating unusual occurrences. If he then wants to protect himself against extreme outcomes among the largest possible observed values, he will call any number as large as 8 or larger an extreme outcome (see area shaded in right hand portion of histogram), because the sum of the probabilities of 8, 9, and 10 is less than .05. On the other hand, if he wishes to protect himself against those extreme outcomes among the smallest possible observed values, we see from the Figure 10–1 that only $n = 0$ and $n = 1$ have small enough combined probabilities to be considered extreme (see area shaded in left hand portion of histogram).

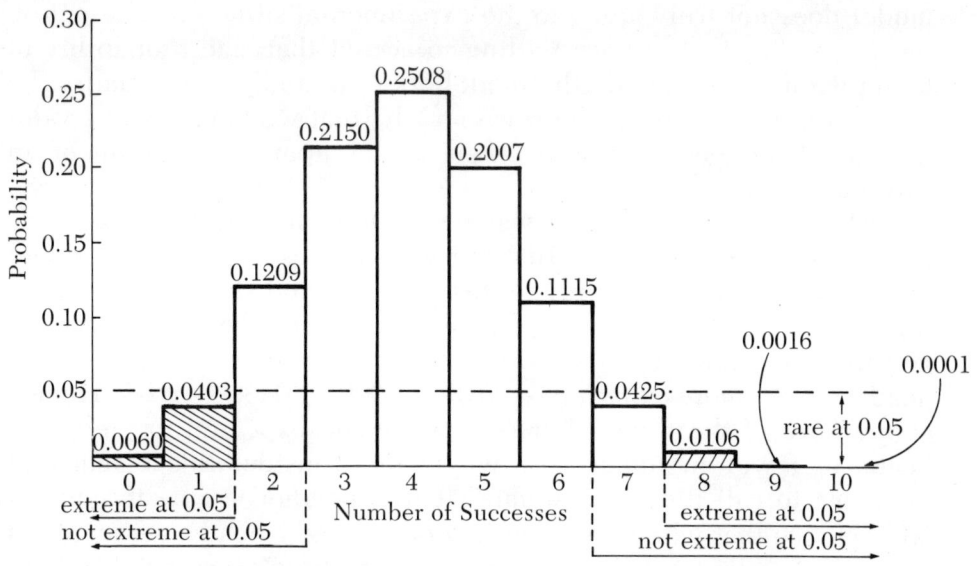

**Figure 10-1** Illustration of extreme and rare events for the binomial model with $N = 10$ and $p = 0.4$.

A rare event would be any event with a probability of 0.05 or smaller. Thus all events with rectangles falling below the dashed line drawn at a height of $p = 0.05$ would be rare events. From Figure 10-1, then, we see that $n = 0$, $n = 1$, $n = 7$, $n = 8$, $n = 9$, and $n = 10$ all yield rare events, while the .05 level extreme events are $n \leq 1$ or $n \geq 8$. The value $n = 7$ is not included as an extreme event since its probability plus those of the events yet more extreme ($n = 8$, $n = 9$, and $n = 10$), add to 0.0548, which is larger than 0.05. We see from this example that not all rare events may be extreme, but all extreme events are rare. Extreme events occur in the "tails" of the distribution, but rare events could conceivably occur anywhere.

Often an experimental procedure involves making a decision as to whether an observation is unusual when a certain chance of success is given by the model. To do so, we may make use of a table of binomial probabilities, such as Table III, to find the probability of observing an outcome of a certain size or one that is even more extreme. Just how extreme an observation must be in order to be unusual is sometimes questionable. As we pointed out before, however, in practice we often use 5% as a "cut-off point." If the event, or one more extreme, can be expected to happen less than 5% of the time, it is considered extreme or unusual.

Thus, when we wish to know whether a certain binomial model in fact applies to a given experimental situation, we may specify the model anticipated by the values of $N$ and $p$ and then perform the experiment. If the outcome of the experiment is not extreme, then we have no reason to doubt the validity of the model. However, if an extreme outcome results, then we have reason to suspect that the

## THE BINOMIAL TEST

model does not truly apply to the experimental situation. The cut-off level specifies that we are willing to accept that size probability of being wrong if we decide the model does not apply. The smaller the level we set, the more certain we may be that we have made a valid decision by rejecting the specified model upon observation of an extreme event.

EXAMPLE 10-1. A coin was tossed ten times and only 2 heads turned up. Is the coin fair? In other words, does the binomial model with $N = 10$ and $p = .5$ apply to this experimental situation? By looking under $N = 10$ and $p = .5$ of Table III, we see that the chances of getting 2 heads or fewer (in other words, an event as extreme as 2 heads or even more extreme) is $.0010 + .0098 + .0439 = .0547$. Hence, about 5.5% of the time a fair coin would be expected to yield two heads or fewer in ten tosses. We would consider this insufficient evidence to call the coin "unfair" if we specified a cut-off value of 0.05. Notice that when the observed result was 2 heads, we were not concerned with finding the chances of getting exactly 2 heads to decide whether this observation was unusual for 10 tosses of a fair coin. Instead we recognized that had we obtained only 1 head or else had we obtained no heads, that would have been even more extreme. Thus, to evaluate whether the outcome was truly unusual, extraordinary, or extreme, we must add the chances of the event observed to those of all events that were even more extreme than that observed. See Figure 10-2 for an illustration of this observation and its implications.

EXAMPLE 10-2. Out of 8 items on an assembly line, 3 were found to be defective. If the probability of a defective is only ten per cent,

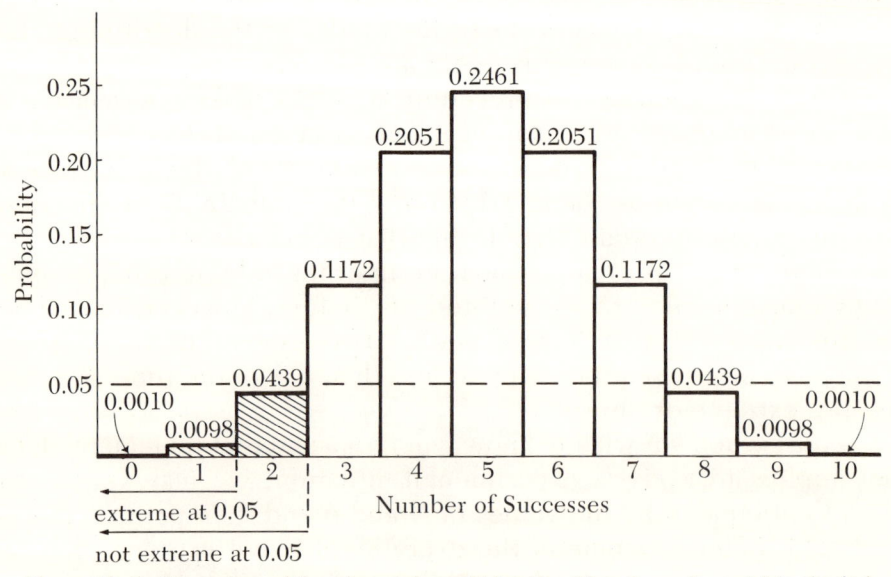

**Figure 10-2** Illustration of the binomial model for $N = 10$ and $p = .5$; $n \leq 2$ is shaded.

## THE BINOMIAL TEST

does production seem to be in control? In Table III, for $N = 8$ and $p = .1$, we see that the chance of getting 3 or more defectives would be $.0331 + .0046 + .0004 + 0 + 0 + 0 = .0381$; that is, less than 4% of the time would we expect 3 or more defectives. Thus, $P = .0381$. We would conclude that this outcome is unusual if production is in control. See Figure 10–3 for an illustration of this extreme event.

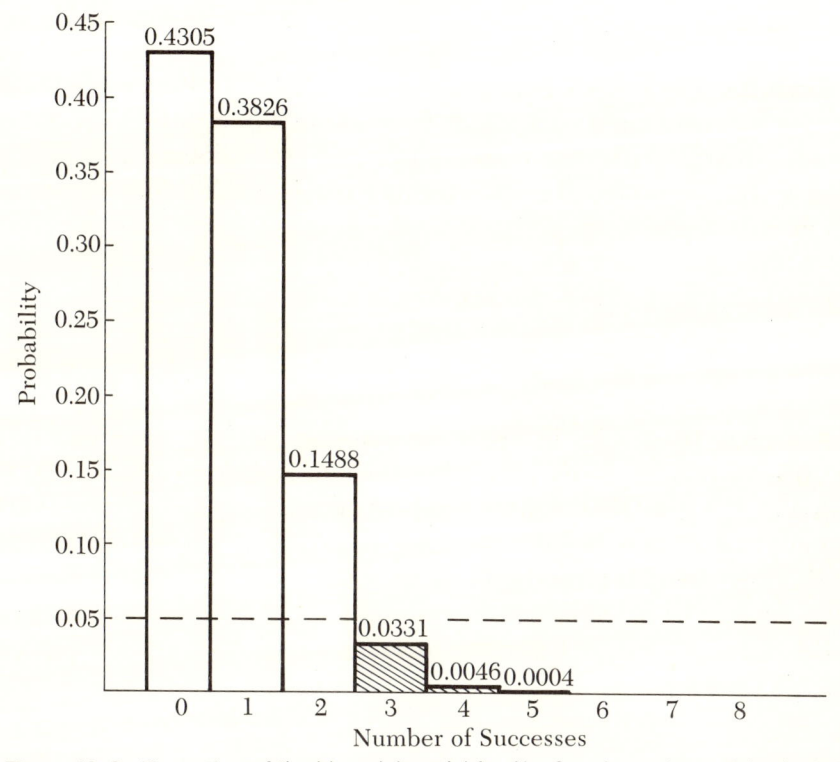

**Figure 10-3** Illustration of the binomial model for $N = 8$ and $p = .1$; $n \geq 3$ is shaded.

What do we do in the case where we have a situation for which we need values of binomial probabilities for $p$ larger than .5? We know that $P(\text{failure}) = 1 - P(\text{success})$; therefore, we need only consider the values in the column given by $1 - P(\text{success})$ and read them "backwards."

EXAMPLE 10–3. Suppose we want to find the chance of getting at least three prizes in 4 boxes of candy if $P(\text{prize}) = .8$ for each box. Since $P(\text{prize}) = .8$, we see that $P(\text{no prize}) = .2$; thus we may read the appropriate probabilities from the $p = .2$ column of Table III if we read them from the opposite end of the column. Since "at least three prizes" means either three or four prizes, it also means zero or one "no-prizes," so from Table III (with $N = 4$), we read the appropriate probability as $P = .4096 + .4096 = .8192$.

## THE BINOMIAL TEST

**Exercise 10-1:** A seed packet stated, "Germination rate: 80%", but when 8 seeds were planted only 2 sprouted. Does this seem to be a case of a misleading advertisement? (Hint: Since $p = .8$ is not in Table III, rephrase the question to read, "Non-germination rate is 20% and 6 did not sprout.")

$P = $ _____; Conclusion: _____

_____

**Exercise 10-2:** An insecticide, advertised as being 90% effective, was sprayed on a sample of 9 cinchbugs and only 4 died. Does this advertising claim seem to be untrue? (Hint: Since $p = .9$ is not in Table III, rephrase the question to read "10% non-effective and 5 cinchbugs did not die.")

$P = $ _____; Conclusion _____

_____

**Exercise 10-3:** In his past experience an instructor had found that 70% of the students in his senior-level class in biology passed the first test. After trying a new method of teaching the subject matter, he found that nine of the ten students in the class passed the first test. Was the new teaching method effective?

$P = $ _____; Conclusion _____

_____

**Exercise 10-4:** If 3 out of 10 college students smoke, would it be unusual if a group of 10 students were chosen at random from the student body and only one of them was a smoker?

$P = $ _____; Conclusion _____

_____

**Exercise 10-5:** In a certain city a survey has shown that 60% of the adult population exhibit some racial prejudices. Out of a sample of 7 pre-school children interviewed, it was found that only 2 showed racial prejudice. Do children tend to be less prejudiced as regards race than their parents?

$P = $ _____; Conclusion _____

_____

# THE BINOMIAL TEST

**Exercise 10-6:** A food pellet is given to a rat each time he turns right at a certain point in a maze, and no reward is given if he turns left. In 9 trials, the rat is observed to turn right 8 times. Has he "learned" his task? (Hint: If there has been no learning, he could be expected to turn right or left at random, so that $p = .5$.)

$P =$ _____; Conclusion _____

_____

**Exercise 10-7:** Toss a coin 10 times and count the number of heads that appear. Is your outcome extreme? _____

**Exercise 10-8:** Toss a die 10 times and count the number of even numbers that appear. Is your outcome extreme? _____

### TERMS TO REMEMBER

**Rare Event:** An event that is unexpected or out of the ordinary; that is, an event that has a small probability of occurrence.

**Extreme Outcome:** An outcome or event whose observed value is an outlier with respect to the value expected under the model.

chapter 11

# THE SIGN TEST

Suppose that the Bestbacon Feed Company has developed a new feeding ration for pigs. They want to test whether this new ration is significantly better than the ration they are currently marketing. They might design an experiment in which one randomly chosen group of pigs is fed with the new ration and one group, also randomly chosen, is fed with the old ration. On the basis of their experimental results, in terms of weight gained by the pigs, they would decide which is better.

The term **factor** denotes a causative agent or an element contributing to the observed effects of treatments. For the example above, the main factor causing weight gain would be the pigs' diet. The term **treatment** means any controlled manipulation of experimental subjects. In some cases it is merely a categorical classification of the subjects based on some factor that the experimenter expects to have an effect on the responses of the subjects. Possible factors to be considered in designing the experiment for the Bestbacon Feed Company would be the sexes of the pigs, their heredity, and their weights prior to being included in the experiment. Different sexes or different breeds might gain weight in different amounts, and larger or smaller initial weights could imply better or worse genetic make-ups for weight gains, even among pigs from the same litter.

Often in experimental work, we encounter what is called a **before-after design;** subjects are measured both before and after treatments to find out the effect of the treatment. This kind of design would not apply to the experiment of the Bestbacon Feed Company which we just discussed, since pigs have to be fed something every day. No valid measurement could be obtained on their overall weight gain by merely measuring them "before" and "after" feeding the new ration, for they would gain weight with any ration as they grew from small pigs to bacon-sized hogs.

Numerous examples of valid before-after designs exist in our everyday experiences. To learn whether a certain teaching method is effective, students may be tested before and after exposure to the teaching method. Production may be measured before and after the inauguration of a new policy regarding coffee-breaks to ascertain whether the new policy results in an improved rate of production. A rat's reaction time could be measured before and after the administration of a stimulant to learn whether the stimulant affected his reaction time.

A similar experimental design results when matched subjects are used for an experiment. Subjects which are chosen because they are alike in some respect and thus are expected to behave similarly under treatment are called **matched subjects.** For example, two leaves from the same tobacco plant might be measured for nicotine content, or five rats from the same litter might be involved in a reaction time experiment. Pupils with the same IQ's might be randomly assigned to one of two experimental classes to find out which of two teaching methods elicits better learning of the subject matter. If the Bestbacon Feed Company had chosen pairs of equal-weight males from the same litters and randomly assigned one of each pair to receive the new and the other to receive the old ration, the result would have been a matched subjects design.

If a before-after design has been used or if matched pairs were involved in an experiment, and if the measurement scale can be assumed to be no stronger than a nominal scale, then a very simple and quick measure of whether or not the treatment had a significant effect can be obtained by using the sign test. The sign test is a form of the binomial test which we considered in Chapter 10. The procedure is very simple. First, we obtain the sign of the difference between the measures for each matched pair. In the case of nominal or ordinal data, we merely record a "+" for a favorable effect due to the treatment, a "−" for an adverse effect, and no sign for no effect. Second, we assume $p = 0.5$ because, if there are truly no treatment effects, there should be an equal number of plus and minus signs which would merely record sampling errors. $N$ is the number of before-after measurements or the number of matched pairs involved, and $n$ is the number of either plus or minus signs, whichever is easier to find. Third, from Table III in the appendix, we find and combine the values of the binomial probabilities corresponding to events as extreme as or more extreme than that which was observed. Finally, we conclude that the treatment had an effect if the sum of the probabilities is less than 5%, indicating

# THE SIGN TEST

**TABLE 11-1** RESULTS OF BEFORE-AFTER TEST.

| Voter | Before | After | Sign of Difference |
|---|---|---|---|
| 1 | for | for | none |
| 2 | against | for | + |
| 3 | against | against | none |
| 4 | for | against | − |
| 5 | for | against | − |
| 6 | against | for | + |
| 7 | against | for | + |
| 8 | against | against | none |
| 9 | against | for | + |
| 10 | against | for | + |
| 11 | for | for | none |
| 12 | against | for | + |

that the outcome is extreme at the 5% level. If the sum exceeds 0.05, we conclude that the treatment had no effect, based on sample evidence.

EXAMPLE 11-1. A local school PTA wanted to know whether it would be worthwhile to purchase TV time to urge voters to pass a school bond issue. They developed a three-minute filmed program that would be used and then obtained the cooperation of a small number of voters who agreed to view the film. Each person's vote was obtained before and after viewing the film and results were as shown in Table 11-1.

For this example, $n = 2$, $p = 0.5$, and $N = 8$, because two minus signs were observed among the 8 matched pairs which had signs. Therefore, the probability of an event as extreme as or more extreme than this, as given by Table III, is $0.1094 + 0.0312 + 0.0039 = 0.1445$; since this is not less than 0.05, it is not considered unusual. From these results we cannot conclude that the filmstrip changed the voters' opinions. Note that the same result would be observed if we had used $n = 6$ for the number of "+" signs instead of $n = 2$ for the number of "−" signs. Recognize that we have imprecise results from small sample size in this instance. The sample evidence does not warrant spending more for the TV advertising without further testing or a better filmstrip.

EXAMPLE 11-2. Seven sets of identical twins were randomly assigned to Groups A and B in such a way that one of each pair was in Group A and one in Group B. The individuals of both groups studied a set of nonsense syllables until each could recall at least 90% of the syllables. Then the members of Group A were given a fairy tale to read, while the members of Group B were given a second set of nonsense syllables to learn. After thirty minutes, both groups were asked to recall the first set of nonsense syllables; the results are shown in Table 11-2.

## THE SIGN TEST

**TABLE 11-2** SCORES OF GROUPS A AND B: PER CENT CORRECT.

|   | Group A | Group B | Group A – Group B Sign |
|---|---------|---------|------------------------|
| 1 | 92 | 85 | + |
| 2 | 83 | 81 | + |
| 3 | 95 | 80 | + |
| 4 | 99 | 93 | + |
| 5 | 90 | 90 | none |
| 6 | 95 | 94 | + |
| 7 | 89 | 87 | + |

Here $N = 6$, $n = 0$, and $p = 0.5$. The probability of observing this result if there were no treatment effect is 0.0156, or less than a 2% chance. Since $n = 0$ is the most extreme event that could occur, we do not have to add probabilities in this case. From the sample evidence, we would conclude that for Group B there was an interference effect when a second set of nonsense syllables had to be learned. This effect was not present for the group that read the fairy tale.

EXAMPLE 11-3. The Underfoot Shoe Company must decide whether to buy Product A or Product B leather conditioner. Both are supposed to result in a deeper, richer color tone in shoes. To test which is best, the company manufactures several pairs of shoes, each pair from the same hide, and treats one of each pair with A and one with B. Then if the color tone of the A-treated shoe is deeper and richer, a "+" is recorded. If that of the B-treated shoe is deeper and richer, a "−" is recorded. If there is no discernible difference, a "0" is recorded. At the conclusion of the experiment, there have been 9 "+" signs, 1 "−" sign, and 4 "0" readings recorded. Thus $N = 10$, $n = 1$, $p = 0.5$, and the probability of this outcome is $0.0098 + 0.0010 = 0.0108$, which is extreme at the 5% level. Based on the experimental evidence the company should use Product A as their leather conditioner.

Notice that the method of analysis in these examples is almost identical to that used in the binomial test of Chapter 10. The only difference is that for the binomial test only one set of sample data was involved, whereas for the sign test two sets of sample data are involved, even though the analysis is performed on the single set of signs of the sample differences. Both the sign test and the binomial test use the binomial distribution of Chapter 9. Tables for binomial probabilities with $N$ larger than ten are available in most libraries, and in Chapter 13 we shall learn a way to compute them.

**Exercise 11-1:** A young forester is being trained to estimate by observation the diameters of trees as he walks through a forest in the company of an experienced forester. They both make note of their

## THE SIGN TEST

estimates of tree diameters. Notice that in this case, the "treatment" is merely a classification of the measurements according to who made them. The results are as follows:

| Tree Number | Young Forester | Trained Forester |
|---|---|---|
| 1 | 6'2" | 5'8" |
| 2 | 4'10" | 4'9" |
| 3 | 3'6" | 3'6" |
| 4 | 4'5" | 4'3" |
| 5 | 5'8" | 5'3" |
| 6 | 6'8" | 6'2" |
| 7 | 5'2" | 5'1" |
| 8 | 4'5" | 4'5" |
| 9 | 3'10" | 3'9" |
| 10 | 5'2" | 5'0" |

Are the young forester's estimates consistently larger or smaller than those of the trained forester?

*Exercise 11-2:* A farmer wishes to test the effectiveness of a new weed control chemical for corn. He lays out ten test plots and carefully marks each test plot into two sections. One section of each plot he cultivates as he normally would, while to the other he applies the weed control chemical. His results from the test plots, measured in bushels yield, are as follows:

| Plot Number | Normally Cultivated | Weed Control Chemical |
|---|---|---|
| 1 | 7.2 | 8.3 |
| 2 | 6.9 | 7.2 |
| 3 | 8.1 | 8.1 |
| 4 | 7.8 | 7.9 |
| 5 | 6.3 | 6.9 |
| 6 | 4.5 | 7.0 |
| 7 | 8.0 | 8.5 |
| 8 | 5.9 | 6.0 |
| 9 | 6.4 | 6.4 |
| 10 | 7.2 | 7.8 |

Do the two methods of weed control result in the same average yield of corn, or does one method appear better than the other in terms of increased yield?

**Exercise 11-3:** The abilities of mental patients to recognize and identify visual objects were recorded before and after electroshock treatments. Do the data indicate that the electroshock treatments had an appreciable effect on the patients' recognition and identification? [Hint: yes-yes and no-no would be tied scores and yes-no would indicate a negative effect, while no-yes would indicate a positive effect.]

| Patient | Before Electroshock | After Electroshock |
|---|---|---|
| 1 | yes | yes |
| 2 | yes | no |
| 3 | no | yes |
| 4 | no | yes |
| 5 | no | no |
| 6 | yes | no |
| 7 | no | no |
| 8 | yes | yes |
| 9 | yes | no |
| 10 | no | yes |

## THE SIGN TEST

**Exercise 11-4:** Two types of egg cartons are being tested by a packaging firm. Ten dozen eggs are subjected, one dozen at a time, to various types of "torture treatments," once in a carton of type A and once in a carton of type B. An egg, once broken, is not available for treatment again. When an egg breaks, the carton could be at fault or the egg could be "weak-shelled." To protect against an untrue conclusion because of weak-shelled eggs, broken eggs are not replaced before transfer of those remaining to the second carton. Therefore, the order of treatment is randomized so that cartons A and B are used first or second randomly. The results are as follows:

| Dozen | A | B |
|---|---|---|
| 1 | X | X |
| 2 | X | W |
| 3 | X | W |
| 4 | W | X |
| 5 | X | W |
| 6 | X | X |
| 7 | X | W |
| 8 | X | W |
| 9 | X | W |
| 10 | X | X |

X means that at least one egg is broken, and W denotes no broken eggs. (Hint: In this case XX or WW would be tied scores or 0's, and WX or XW would indicate one or the other of the cartons is better. Either WX or XW could be denoted "+" with the other denoted "−".)

Are the cartons equally good, or is one better than the other in terms of keeping the eggs unbroken?

**Exercise 11-5:** Hold a die in each hand and toss both dice together. Repeat the experiment ten times. Record the numbers carefully, according to whether the die was tossed by the left or right hand, in the table below. From your experimental data, does your right hand tend to toss larger numbers? Repeat the experiment, tossing the dice in an unbiased manner and using your favorite lucky charm to try to get larger numbers on the right-handed toss.

## THE SIGN TEST

|       | Trial without Charm |       |      | Trial with Charm |       |      |
| :---: | :---: | :---: | :---: | :---: | :---: | :---: |
| Trial | LEFT | RIGHT | SIGN | LEFT | RIGHT | SIGN |
| 1 | | | | | | |
| 2 | | | | | | |
| 3 | | | | | | |
| 4 | | | | | | |
| 5 | | | | | | |
| 6 | | | | | | |
| 7 | | | | | | |
| 8 | | | | | | |
| 9 | | | | | | |
| 10 | | | | | | |

Answer _____    Answer _____

_____    _____

Did your "luck" change? _____

**Exercise 11-6:** Toss one coin with the left hand and one with the right hand at the same time. Repeat ten times. Record a "match" (HH or TT) as + and a "nonmatch" (HT or TH) as −. Do your hands tend to "work together" to give matches? Repeat the experiment, tossing the coins in an unbiased manner and using your favorite lucky charm to obtain more matches.

|       | Trial without Charm |       |      | Trial with Charm |       |      |
| :---: | :---: | :---: | :---: | :---: | :---: | :---: |
| Toss | RIGHT | LEFT | SIGN | RIGHT | LEFT | SIGN |
| 1 | | | | | | |
| 2 | | | | | | |
| 3 | | | | | | |
| 4 | | | | | | |
| 5 | | | | | | |
| 6 | | | | | | |
| 7 | | | | | | |
| 8 | | | | | | |
| 9 | | | | | | |
| 10 | | | | | | |

## THE SIGN TEST

Answer _____                    Answer _____

        _____                                         _____

Did your "luck" change? _____

### TERMS TO REMEMBER

**Factor:** A causative agent; an element contributing to the observed effects of treatments.

**Treatment:** Controlled manipulation or categorical classification of subjects.

**Before-After Design:** An experiment designed in such a way that the subject is measured before and after treatment in order to assess the effect of the treatment.

**Matched Subjects:** Subjects from two or more treatment groups that have been chosen in such a way that their responses are expected to be similar because of the criterion upon which they were matched.

chapter 12

# SIGNED RANKS TEST

In the last few chapters we have been concerned with the probabilities of occurrence of experimental outcomes, some of which were extreme outcomes. We saw that any experimental outcome may be attributed to chance, but extreme outcomes which could be expected to occur very seldom are unlikely to arise by chance. When these extreme outcomes occur, they usually cause the experimenter to suspect that the response is due to the treatment applied or to some factor other than chance fluctuations, even though he recognizes that they could have occurred by chance.

Often when analyzing sample data, the procedure calls for calculating a numerical quantity from the sample data. This quantity is then compared to a tabled value which identifies an extreme outcome. The quantity calculated from sample data is called a **statistic**, and the tabled value indicating an extreme outcome is called a **critical value** or **critical point.** Thus the mean, or the mode, or the number of plus signs for a sign test are all statistics; the values in Table II indicating the upper and lower cut-off values for the total number of runs in a random sequence are critical points. These critical values, usually found in convenient tables, are merely cut-off points telling how large is "too large" or how small is "too small." They indicate the values which would be expected to occur by chance only a very small proportion of the time.

The numbers "47" and "49" are different; anyone would agree

# SIGNED RANKS TEST

that they are not equal! But suppose these numbers were among the observations 2, 3, 17, 19, 27, 39, 47, 49, 82, 96, 108, 125. Now they don't seem quite so different. On the other hand, if they appeared among the observations 47, 47.3, 47.5, 48, 48.4, 48.9, 48.9, 49, they would seem quite different. Similarly, when analyzing experimental results, the differences present among certain values are not easy to discern. Then we need tables of critical values to make the decisions for us. Sometimes, as with Table II, we obtain critical points giving values of extreme observations. Other times, as with Table III, we calculate the probability of observing a certain outcome and then decide whether the outcome was extreme.

Suppose that we wanted to analyze data from an experiment which had a before-after design or else a design using matched subjects. Suppose further that the measurement scale involved was ordinal. The only difference between the design of this experiment and the type of experiment considered in Chapter 11 is the measurement scale. Whereas the sign test considers only the positive or negative *direction* of the differences, the Wilcoxon signed ranks test also takes into account the ranked *magnitudes* of the differences.

The sign test may be used as a "quick and dirty" test anytime that matched pairs or a before-after design is involved. However, it is appropriately used only when the measurement scale is nominal. The signed ranks test may be used when the measurement is at least ordinal. When used with the data having the ordinal measurement scale, the signed ranks test will be more precise than the sign test. This increased precision means that true differences between the two sample means will be detected more often with the signed ranks test than will be the case with the sign test. In other words, though the sign test is "quick" in the sense of being very easy to perform, it is "dirty" in the sense that other tests may be more precise.

The experimenter should first choose the test he wishes to use and then gather the sample data. The test he chooses should be the one which best suits his specifications for the design of his experiment, the measurement scale involved, and the precision desired in the result relative to the time available for the analysis.

EXAMPLE 12-1. In Example 11-1, we were evaluating the effect of the PTA's filmstrip on voter preference. There was no way to measure magnitude of preference, because everyone was either "for" or "against" the bond issue. Thus, we could not use a signed ranks test for these data. However, in Example 11-2, which concerned the sets of identical twins who had learned nonsense syllables, we could decide whether the measures of location differ by using a signed ranks test. The per cent correct for each group, the signs of the differences, and the ranks of the differences are shown in Table 12-1.

The actual percentages recorded would appear to be in the interval measurement scale. However, the experimenter may realize

## SIGNED RANKS TEST

**TABLE 12-1** DATA FOR EXAMPLE 12-1.

| (1)<br>Twin Pair<br>Number | (2)<br>Group A | (3)<br>Group B | (4)<br>Sign | (5)<br>Difference | (6)<br>Rank |
|---|---|---|---|---|---|
| 1 | 92 | 85 | + | +7 | +5 |
| 2 | 83 | 81 | + | +2 | +2.5 |
| 3 | 95 | 80 | + | +15 | +6 |
| 4 | 99 | 93 | + | +6 | +4 |
| 5 | 90 | 90 | none | — | — |
| 6 | 95 | 94 | + | +1 | +1 |
| 7 | 89 | 87 | + | +2 | +2.5 |

Sum of positive ranks = 21        Sum of negative ranks = 0

that these values are not really so precise as they appear—perhaps because of poor lighting in the room where the test was taken or because of noise or confusion during the testing period. He may prefer to consider them as having only ordinal strength. For example, the difference of "7" for twin pair number 1 would be considered larger than the "6" for twin pair number 4, but less than the "15" for twin pair number 3. Unlike the sign test for which all three of these measurements are merely "+'s", the signed ranks test will take into consideration these differences in rank sizes, although it will not differentiate between actual sizes of the differences in the numerical scores.

To perform the signed ranks test we proceed as follows:
(1) Calculate the sizes of the differences as is done in column 5.
(2) Assign ranks to the differences in ascending order of their numerical sizes regardless of the signs, as is done in column 6. In case of tied differences, get the average rank—such as the rank of 2.5 in place of the ranks 2 and 3, given to the differences of "2" for both twin pair number 2 and twin pair number 7.
(3) Assign to the ranks the same signs that the differences had. In the above example all the differences were + or zero, so that all the ranks in column 6 have a + sign; the zero is ignored as it was with the sign test.
(4) Calculate the statistic $T$ = the numerical value of the total of the positive ranks or of the negative ranks, whichever is smaller. In the example $T = 0$ since there are no negative ranks.
(5) Find the probability of a $T$ that size or smaller in Table IV of the appendix. In this example we see that for a sample size of 6, a value of $T = 0$ would occur by chance between 2% and 5% of the time if there were in fact no true differences between Groups A and B.

## SIGNED RANKS TEST

**TABLE 12-2** DATA FOR EXAMPLE 12-2.

| Tangerine No. | 1 | 2 | 3 | 4 | 5 | 6 | 7 | 8 | 9 | 10 | 11 | 12 | 13 |
|---|---|---|---|---|---|---|---|---|---|---|---|---|---|
| Sunny Side | 7.6 | 8.2 | 7.7 | 7.9 | 8.6 | 8.9 | 10.1 | 9.1 | 6.5 | 7.8 | 8.4 | 8.1 | 9.5 |
| Shady Side | 6.9 | 8.4 | 7.1 | 7.7 | 8.7 | 8.9 | 10.2 | 9.4 | 6.2 | 7.3 | 8.0 | 8.2 | 9.3 |
| Difference | +.7 | −.2 | +.6 | +.2 | −.1 | 0.0 | −.1 | −.3 | +.3 | +.5 | +.4 | −.1 | +.2 |
| Unsigned Ranks | 12 | 5 | 11 | 5 | 2 | — | 2 | 7.5 | 7.5 | 10 | 9 | 2 | 5 |
| Signed Ranks | +12 | −5 | +11 | +5 | −2 | — | −2 | −7.5 | +7.5 | +10 | +9 | −2 | +5 |

Sum of positive ranks = 59.5   Sum of negative ranks = 18.5

Table IV lists the values of the calculated statistic $T$ which would be considered extreme in the sense that if the treatment had no effect, then a value that size or smaller would occur less than 1%, or 2%, or 5%, or 10% of the time (according to which column of the table is read). Thus, for a given tabled value of $T$, any value that size or smaller would be considered extreme at the given probability level.

EXAMPLE 12-2. Suppose that we measure the percentages of solids in the two halves of each of 13 tangerines. One of the halves is the "outside half," which would get sunlight, and the other is the "inside half," which would be shaded by the tree. The question is whether the sunlight causes a difference in the percentage of solids between the two halves. The data might appear as in Table 12-2.

Notice that no rank is assigned to the zero difference. Notice also that when the ranks are assigned, the signs of the differences are ignored. Tied ranks are averaged: the two +.2's and the one −.2 which would normally rank 4th, 5th, and 6th are all assigned rank 5, since $\frac{4+5+6}{3} = 5$. Then the ranks are given the same signs as the corresponding differences have. In this case, because the sum of the negative ranks is the smaller, it is called $T$; that is, $T = 18.5$. Also, $N = 12$, because one zero was ignored. From Table IV we observe that a value of $T = 17$ or less has a probability of 10% of occurring if there is no true difference. Hence, we conclude that a $T$-value of 18.5 or less would occur more than 10% of the time if there were truly no difference in the percentages of solids in the two halves of the tangerines. Consequently, our sample data would cause us to believe that there is no true difference between the percentages of solids contained in the shaded and sunlit halves of the tangerines.

In the first two exercises below we shall find out whether your right hand knows what your left hand is doing. The remainder of the exercises are designed to provide practice in performing a signed-ranks test.

***Exercise 12-1:*** Take two dice, one in each hand, and toss them, noting carefully which die came from which hand. Repeat the experiment 14 times, for a total of 15 tosses. The object is to see whether the die in your right hand tends to come up consistently with a larger or a smaller number than does that in your left hand. (Thus we are assuming right-left to be a matching or pairing criterion—is it?) Record your results in the following table and decide whether there is a true difference between left and right by using a signed ranks test.

|  | Toss Number | 1 | 2 | 3 | 4 | 5 | 6 | 7 | 8 | 9 | 10 | 11 | 12 | 13 | 14 | 15 |
|---|---|---|---|---|---|---|---|---|---|---|---|---|---|---|---|---|
| Number on Die | Right Hand | | | | | | | | | | | | | | | |
|  | Left Hand | | | | | | | | | | | | | | | |

**Exercise 12-2:** Open your local phone book to any page. Point out a starting place with your right hand and another with your left hand. Record the digits of five phone numbers for each hand, one at a time, using only the last three digits of any single phone number, until you have 15 observations for each hand. For example, if the entries read 784-7731, 784-9462, 784-9483, you would record your first nine numbers for that hand as 7-3-1-4-6-2-4-8-3. Record your data observations in the following table and test whether the observations differ for the right and left hand by using a signed ranks test.

|  | Digit Number | 1 | 2 | 3 | 4 | 5 | 6 | 7 | 8 | 9 | 10 | 11 | 12 | 13 | 14 | 15 |
|---|---|---|---|---|---|---|---|---|---|---|---|---|---|---|---|---|
| Size of Digit | Right Hand | | | | | | | | | | | | | | | |
|  | Left Hand | | | | | | | | | | | | | | | |

**Exercise 12-3:** Redo Exercise 11-1 using a signed-ranks test. Do your results differ from what you obtained in Chapter 11?

## SIGNED RANKS TEST

***Exercise 12-4:*** Redo Exercise 11-2 using a signed-ranks test. Do your results differ from what you obtained in Chapter 11?

### TERMS TO REMEMBER

**Statistic:** A quantity calculated from sample data.

**Critical Value:** The largest (or smallest, according to the test involved) value of a calculated statistic which would indicate an extreme outcome if the model assumed were true.

**Critical Point:** Another name for critical value.

chapter 13

# PERMUTATIONS AND COMBINATIONS

Before considering permutations, we must become familiar with a bit of mathematical shorthand. The **factorial notation,** $n!$, means the product of the first $n$ positive consecutive integers, or $n \cdot (n-1) \cdot (n-2) \cdot \ldots \cdot 2 \cdot 1$. For example, $3! = 3 \cdot 2 \cdot 1 = 6$; and $5! = 5 \cdot 4 \cdot 3 \cdot 2 \cdot 1 = 120$. The results become large at an alarming rate—hence the exclamation point. The value of $12!$ is 479,001,600. Obviously $1! = 1$, and $0!$ is by definition also equal to 1.

**Exercise 13-1:** Calculate the following factorials.

$4! = $ _____

$6! = $ _____

$(7-3)! = $ _____

$7! - 3! = $ _____

$\dfrac{8!}{6!} = $ _____

$\dfrac{8!}{5!3!} = $ _____

$\dfrac{3!}{0!} = $ _____

**Exercise 13-2:** Calculate the following factorials.

$2! = $ _____

## PERMUTATIONS AND COMBINATIONS

$7! = $ _____

$(6-2)! = $ _____

$6! - 2! = $ _____

$\dfrac{9!}{7!} = $ _____

$\dfrac{9!}{6!3!} = $ _____

$\dfrac{4!}{0!} = $ _____

Now that we are familiar with the factorial notation, we are ready to investigate methods of counting that involve factorials. Suppose we are interested in knowing:
1. How many batting orders are possible with 9 players.
2. How many ways first, second, and third places may be awarded in a race with twelve entrants.
3. How many different seating arrangements are possible with 8 people in a row on a stage.
4. How many ways 3 cars can pull into 3 parking places.
5. How many three-digit numbers less than 600 can be formed using the digits 1,2,3,4,5,6.
6. How many ways 6 children can form 2-person "teams" to wash and dry dishes.

Notice that one common feature of all of these questions is ORDER. For number 1, two different batting orders may use the same players, but they differ because of the order in which the players come to bat. For number 4, Jones in first place, Smith in second, and Kennedy in third is a different arrangement from Jones second, Smith first, and Kennedy third. For number 5, 123 is not the same number as 321. In other words, in all of these questions, the order of the item is important.

If there are only a few items to be ordered, it is easy to list all possible orderings. For example, the different orderings, or **permutations** of the letters A, B, and C are ABC, ACB, BAC, BCA, CAB, and CBA. However, when a large number of objects is involved, such as the possible orderings of the first twelve letters of the alphabet, this process becomes tedious. Furthermore, it is unnecessary if one merely wants to know the total number of possible orderings.

The **counting rule** is the most general method of counting that we shall study in this chapter. It may be used whenever one wishes to count the number of ways that several things can be done simultaneously. The counting rule tells us that if one thing can be done in $n_1$ ways and a second in $n_2$ ways, then both things can be done together in $n_1 n_2$ ways. If a third thing can be done in $n_3$ ways, then all three things can be done together in $n_1 \cdot n_2 \cdot n_3$ ways; and for as many more different things as are involved, we merely multiply by the number of ways each can be done.

To solve each problem we may first ask: "How many different things are to be done?" and then draw that many blanks on the paper. Next we ask, "How many ways can the first thing be done?" and then put that number in the first blank. We continue by asking, "How many ways can the second thing be done?" and then put that number in the second blank. When all blanks are filled, the numbers in them are then multiplied together. We may use this counting rule to calculate the total number of ways that the first twelve letters of the alphabet may be ordered: $\underline{12} \cdot \underline{11} \cdot \underline{10} \cdot \underline{9} \cdot \underline{8} \cdot \underline{7} \cdot \underline{6} \cdot \underline{5} \cdot \underline{4} \cdot \underline{3} \cdot \underline{2} \cdot \underline{1}$ or 12!, which as we saw before equals $\underline{479,001,600}$ different ways.

The reason for multiplying the numbers of ways together is obvious when a "tree" diagram is drawn to illustrate the problem. Let us turn to the simple problem of the different orderings of the letters A, B, and C to illustrate this point. If we want to know how many ways these letters may be ordered, we note that any one of three letters, A, B, or C, may go in the first position. Then for each of those three possibilities two letters remain for the second position. Finally, there is only one letter remaining for the third position, as outlined by the tree diagram in Figure 13–1. Notice that after one letter has been specified for the first position, only two are left for the second position —in other words, repetition of a letter is not possible when we consider the orderings of the letters. Hence, we do *not* have $3 \cdot 3 \cdot 3$ or 27 ways; we have $3 \cdot 2 \cdot 1$ ways, which is defined as $3! = 6$.

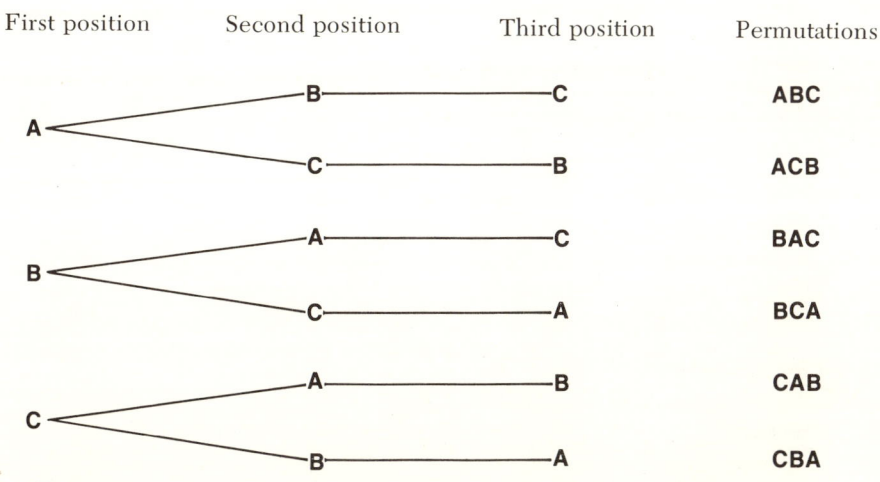

**Figure 13–1** Tree diagram for the number of permutations of the letters A, B, C.

## PERMUTATIONS AND COMBINATIONS

In summary, if we want to know the number of ways that we may **permute**, or reorder, the letters A, B, and C, we imagine that there are three blanks to be filled, one corresponding to each position for a letter. We fill these imaginary blanks with the number representing the number of letters that can possibly be put in that position, remembering that a letter once used is not available for use again. Then we multiply these numbers to find the total number of permutations: $\underline{3} \cdot \underline{2} \cdot \underline{1} = 3! = 6$, by the counting rule. These concepts are further illustrated by the next two simple examples.

EXAMPLE 13-1. Mary has 4 skirts and 5 sweaters. Since they all are different colors, order counts. How many different outfits may she

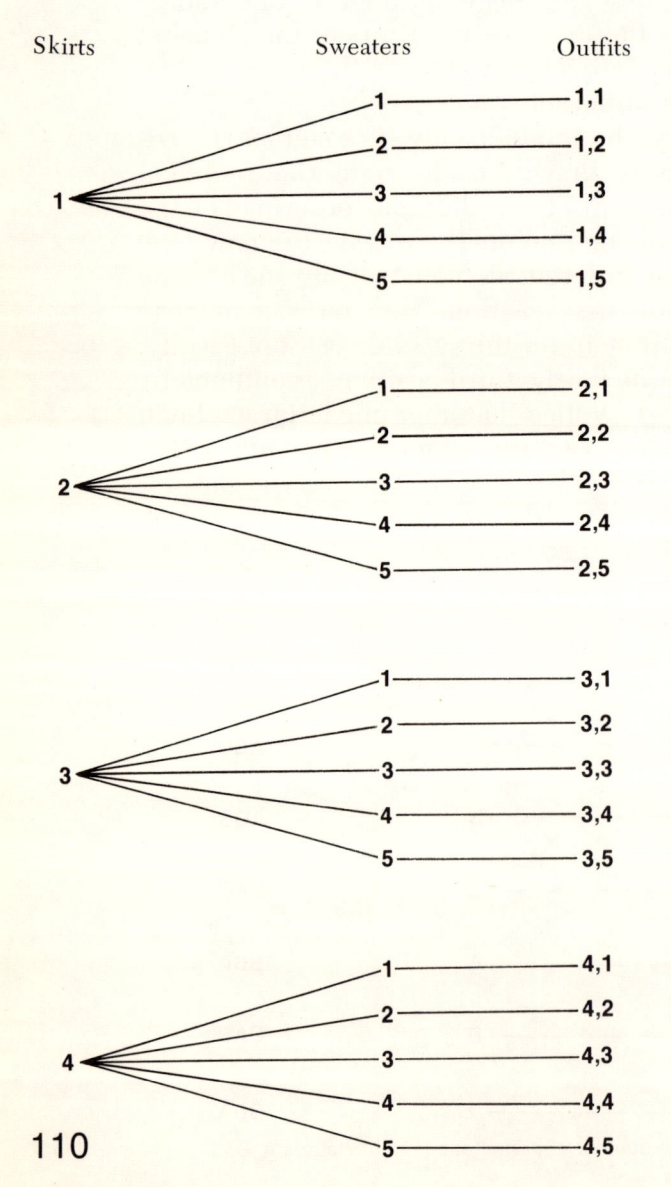

**Figure 13-2** Tree diagram illustrating different possible outfits.

# PERMUTATIONS AND COMBINATIONS

have if any skirt may be worn with any sweater? (See Figure 13-2.) $\underline{4} \cdot \underline{5} = 20$.

EXAMPLE 13-2. Two coins are tossed. How many different outcomes are possible? (See Figure 13-3.) $\underline{2} \cdot \underline{2} = 4$.

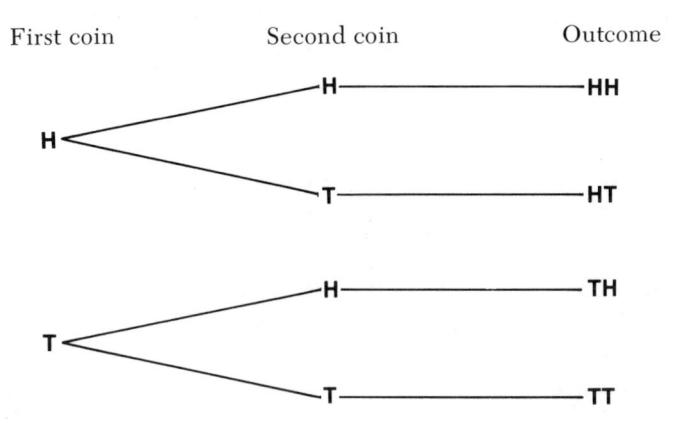

**Figure 13-3** Tree diagram illustrating possible outcomes for the toss of two coins.

Not always will there be as many spots to fill as there are objects to fill them. For example, suppose that there are twenty available dance dates on a university calendar, and thirty student organizations each want to sign up for one dance date. In this situation there are twenty blanks to fill, and we can fill the first with any one of the thirty names of groups, the second with any one of the remaining twenty-nine, and so on. Finally, after the first nineteen blanks are filled, there are eleven names left for the twentieth blank. The result is then, by the counting rule, $30 \cdot 29 \cdot 28 \cdot 27 \cdot 26 \cdot 25 \cdot 24 \cdot 23 \cdot 22 \cdot 21 \cdot 20 \cdot 19 \cdot 18 \cdot 17 \cdot 16 \cdot 15 \cdot 14 \cdot 13 \cdot 12 \cdot 11$, which, when multiplied, is one of those alarmingly large numbers! It is much easier to express the result as $\frac{30!}{10!}$. This kind of result is usually left in factorial form unless we have computers available for our use. Thus, the general method of calculating the number of permutations of any $n$ things taken only $r$ at a time is $P(n,r) = \frac{n!}{(n-r)!}$. (Note: Do not confuse this notation, in which the $P$ stands for "permutation," with the notation in which $P$ stands for "probability." It will usually be clear which is meant.) In the example we just discussed, $n = 30$ student organizations and $r = 20$ dance dates were specified, so that we were considering all different possible permutations or orderings of 30 groups taken 20 at a time. We arrived at a final result of $\frac{30!}{10!}$, which could be expressed $\frac{30!}{(30-20)!}$. The method of permutations is merely a special case of the

**PERMUTATIONS AND COMBINATIONS**

counting rule: the "number of ways" of doing something is simply all of the different possible orderings of a fixed number of different things in a specified number of positions. In each of the two simple examples following we shall observe that these problems can be solved equally well by using the counting rule or by using the permutations formula.

EXAMPLE 13-3. Suppose we are playing "musical chairs" and there are 5 people to play and only 3 chairs. How many ways can the people be seated? To begin with, there are only three blanks to fill. Thus $\underline{5} \cdot \underline{4} \cdot \underline{3} = 60$. Our solution was done by the counting rule! But here we have the special case for which we were permuting 5 people in 3 chairs, and so we could have solved it just as easily by using $P(5,3) = \frac{5!}{2!} = \frac{5 \cdot 4 \cdot 3 \cdot 2 \cdot 1}{2 \cdot 1} = 5 \cdot 4 \cdot 3 = 60.$

EXAMPLE 13-4. Suppose we went to a sale and bought 6 cans of paint, all of different colors. We want to repaint two rooms. How many different color combinations could we have if we wanted to paint each room a single color and intended to paint each room a different color? $\underline{6} \cdot \underline{5} = P(6,2) = \frac{6!}{4!} = 30.$

**Exercise 13-3:** Answer questions 1 through 3 at the beginning of this chapter (page 108).

1. _____

2. _____

3. _____

**Exercise 13-4:** Answer questions 4 through 6 at the beginning of this chapter (page 108).

4. _____

5. _____

6. _____

Suppose next that we are interested in finding the number of ways:
7. a committee of 5 may be chosen in a club with 25 members;
8. ten beads can be drawn from an urn containing 1000 identical beads;
9. 2 red roses and 3 white roses can be chosen to form a corsage if there are 7 identical red roses and 10 identical white roses to choose from;
10. ten identical balls can fall into 3 unordered slots;

## PERMUTATIONS AND COMBINATIONS

11. exactly one head can appear when a coin is tossed three times;
12. a committee of 3 Democrats and 2 Republicans may be chosen from a Senate group which has a total of 57 Democrats and 43 Republicans.

Notice that in each case, order is not important. A committee is the same committee regardless of the order in which its members' names are read. If one is interested only in the number of heads, then the order in which they appear (i.e., first, second, or third) does not matter. And Democrats remain Democrats regardless of their seating arrangement at a committee table. In this case, as in the case of permutations, we are in effect considering a smaller group of $r$ things taken from a larger group of $n$ things. In contrast to permutations, however, the **combinations** of the $n$ things taken $r$ at a time is the number of ways one may select $r$ things from a larger group of $n$ things *if the order of their arrangement is irrelevant.*

The basic difference between finding the combinations of $n$ things taken $r$ at a time and finding the permutations of $n$ things taken $r$ at a time is simply the question of whether the order of the groupings makes a difference. Therefore, we may first obtain the number of permutations of the $n$ things taken $r$ at a time, and then *remove the effect of order.* In other words, if we want to know how many groups of 3 letters each may be chosen from the letters ABCDEF, we may begin by calculating the number of permutations of the six things taken three at a time: $6 \cdot 5 \cdot 4$ or $\frac{6!}{3!}$. Now recognize that we only need to remove from this number the effect of order. These 120 ways include, for example, ABC, ACB, BAC, BCA, CAB, and CBA, all six of which form the same group of 3 letters. If we divide the expression $\frac{6!}{3!}$ by the number of ways the 3 things can be permuted among themselves, namely 3!, the result, $\frac{6!}{3!3!} = \frac{6 \cdot 5 \cdot 4}{3 \cdot 2 \cdot 1} = 20$, is the total number of different groups of letters with the order of letters ignored. We have derived the formula used to calculate the number of combinations of $n$ things taken $r$ at a time, namely $C(n,r) = \frac{n!}{(n-r)!\,r!}$. From the derivation, we see that $C(n,r) = \frac{P(n,r)}{P(r,r)}$.

EXAMPLE 13–5. In how many ways can a group of 3 children be chosen from a classroom of 30 children? Here, the order of their appearance is immaterial. We can, however, proceed again from the basic counting formula. If order made a difference, the result would be $30 \cdot 29 \cdot 28$. But since order does not count, we need to divide by the number of ways the 3 children may be ordered, namely $3 \cdot 2 \cdot 1$. Thus we have $\frac{30 \cdot 29 \cdot 28}{3 \cdot 2 \cdot 1} = C(30,3) = \frac{30!}{27!\,3!} = 4060$.

## PERMUTATIONS AND COMBINATIONS

EXAMPLE 13–6. How many permutations are possible of the letters in the word "STATISTICS"? Using the term "permutations" implies that order counts, but we cannot distinguish among the 3 S's, and the 2 I's, and the 3 T's. Therefore, for these particular letters, we must remove the effect of order, just as we would for combinations. To do so we divide by the number of permutations of three letters, that of two letters, and that of another three letters. Since there are a total of 10 letters in the word, we have $\frac{10!}{3!2!3!} = \frac{10 \cdot 9 \cdot 8 \cdot 7 \cdot 6 \cdot 5 \cdot 4 \cdot 3 \cdot 2 \cdot 1}{3 \cdot 2 \cdot 1 \cdot 2 \cdot 1 \cdot 3 \cdot 2 \cdot 1} =$ a large number again!

The three basic counting methods which we have investigated in this chapter are the counting rule, the permutations of $n$ things taken $r$ at a time, and the combinations of $n$ things taken $r$ at a time. In some cases different counting methods may be combined in one problem. Then the problem must be broken down into its component parts for solution. The parts will be added or multiplied together. Two simple rules of thumb are:

(1) If the wording of the problem implies *"and,"* then *multiply*.
(2) If the wording implies *"or,"* then *add*.

Notice the use of these rules in Examples 13–7 and 13–8.

EXAMPLE 13–7. In how many ways may numbers of at most 3 digits each be formed from the digits 1, 2, 3, 4, 5, and 6, if repetition of digits is not allowed?

To begin with, we must realize that "at most 3" means 3 *or* 2 *or* 1. Thus, the number of different orderings of digits must be found for 3-digit numbers, 2-digit numbers, and 1-digit numbers separately, and the results *added*. Since, for example, the 3-digit number has a first *and* a second *and* a third digit, this sub-result could be obtained by the counting formula, namely $\underline{6} \cdot \underline{5} \cdot \underline{4}$; the same can be done for the 2-digit and 1-digit numbers. However, since repetition is not allowed and order counts, the solution to the problem may also be obtained using the permutations formula. Thus we have

$$P(6,3) \quad + \quad P(6,2) \quad + \quad P(6,1) \quad =$$

$$\frac{6!}{3!} \quad + \quad \frac{6!}{4!} \quad + \quad \frac{6!}{5!} \quad =$$

$$\underline{6} \cdot \underline{5} \cdot \underline{4} \quad + \quad \underline{6} \cdot \underline{5} \quad + \quad \underline{6} \quad =$$

$$\underbrace{120}_{\text{The number of 3-digit numbers}} + \underbrace{30}_{\text{The number of 2-digit numbers}} + \underbrace{6}_{\text{The number of 1-digit numbers}} = 156$$

Results such as this may be left in factorial form if desired, especially when they are large numbers.

**PERMUTATIONS AND COMBINATIONS**

EXAMPLE 13-8. Among those present at a convention of university professors, there are 25 from New York, 15 from Michigan, and 18 from California. A committee is to be chosen, consisting of 2 from New York, 1 from Michigan, and 2 from California. How many different committees are possible? Since on a committee the order of the members is irrelevant, we have a combinations problem. We wish to choose 2 out of 25 from New York *and* 1 out of 15 from Michigan *and* 2 out of 18 from California. Thus, our solution would be

$$C(25,2) \cdot C(15,1) \cdot C(18,2) = \frac{25 \cdot 24}{2 \cdot 1} \cdot \frac{15}{1} \cdot \frac{18 \cdot 17}{2 \cdot 1}.$$

*Exercise 13-5:* Answer questions 7 through 9 following Exercise 13-4.

7. _____

8. _____

9. _____

*Exercise 13-6:* Answer questions 10 through 12 following Exercise 13-4.

10. _____

11. _____

12. _____

*Exercise 13-7:* Make a tree diagram showing the total number of outcomes when a red die and a green die are tossed.

## PERMUTATIONS AND COMBINATIONS

Use the counting formula to find out how many possibilities there are.

---

***Exercise 13-8:*** Make a tree diagram showing the total number of outcomes when a coin and a die are tossed together.

Use the counting formula to find out how many possibilities there are.

---

***Exercise 13-9:*** A contractor is prepared to build houses with 2, 3, or 4 bedrooms; with or without a basement; and with one and one-half, 2, or 3 bathrooms. Make a tree diagram showing the total number of model houses that would have to be built by the contractor to show the public all possible plans.

**PERMUTATIONS AND COMBINATIONS**

Use the counting formula to find out how many possibilities there are.

_____

**Exercise 13-10:** Make a tree diagram showing the total number of ways you could order lunch if you were to choose your sandwich from a hamburger, a cheeseburger, or a bacon-lettuce-tomato; and if you were to choose your drink from a milk shake, a soda, or an orange freeze; and if you were to choose either french fries or onion rings.

Use the counting formula to find out how many possibilities there are.

_____

**Exercise 13-11:** Find the number of combinations of 3 things taken 0 at a time (in other words, out of 3 objects, none will be chosen) _____; 3 things taken 1 at a time _____; 3 things taken 2 at a time _____; 3 things taken 3 at a time _____. Compare your answers with the third line of the Pascal Triangle. _____

_____

**Exercise 13-12:** Find the number of combinations of 4 things taken 0 at a time (in other words, out of 4 objects, none will be chosen) _____; 4 things taken 1 at a time _____; 4 things taken 2 at a time _____; 4 things taken 3 at a

**PERMUTATIONS AND COMBINATIONS**

time _____; 4 things taken 4 at a time _____. Compare your answers with the fourth line of the Pascal Triangle. _____

_____

**Exercise 13-13:** Show that C(3,1) = C(3,2), that C(4,4) = C(4,0), and that C(5,2) = C(5,3).

Do you observe the same thing in the Pascal Triangle? _____

What would be the statement of the general rule? _____

_____

**Exercise 13-14:** Show that C(3,0) = C(3,3), that C(4,3) = C(4,1), and that C(5,1) = C(5,4).

Do you observe the same thing in the Pascal Triangle? _____

What would be the statement of the general rule? _____

**PERMUTATIONS AND COMBINATIONS**

**TABLE 13-1** ILLUSTRATION OF THE RELATIONSHIP BETWEEN THE $(r+1)$st ENTRY OF THE $n$th ROW OF THE PASCAL TRIANGLE, AND $C(n,r)$. THE FIRST DIAGONAL COLUMN OF THE TRIANGLE CORRESPONDS TO THE FIRST DIAGONAL COLUMN OF $C(n,r)$, THE SECOND TO THE SECOND, AND SO ON, AS INDICATED BY THE DASHED LINES.

| $n$ | Pascal Triangle | $C(n,r)$ |
|---|---|---|
|   |   | $r=0$ $\quad$ $r=1$ $\quad$ $r=2$ $\quad$ $r=3$ $\quad$ $r=4$ |
| 1 | 1  1 | $C(1,0)$ $\;$ $C(1,1)$ |
| 2 | 1  2  1 | $C(2,0)$ $\;$ $C(2,1)$ $\;$ $C(2,2)$ |
| 3 | 1  3  3  1 | $C(3,0)$ $\;$ $C(3,1)$ $\;$ $C(3,2)$ $\;$ $C(3,3)$ |
| 4 | 1  4  6  4  1 | $C(4,0)$ $\;$ $C(4,1)$ $\;$ $C(4,2)$ $\;$ $C(4,3)$ $\;$ $C(4,4)$ |

Exercises 13-11 through 13-14 illustrate the fact that the entries in the Pascal Triangle are in fact the number of combinations of $n$ things taken $r$ at a time. Table 13-1 shows the correspondence between the triangle and combinations.

Thus we see that if $n$ is small, the easiest way to find the number of combinations is by quickly developing a small Pascal Triangle having $n$ rows; but if $n$ is large, then it is easier to obtain $C(n,r)$ directly.

**TERMS TO REMEMBER**

**Permutations:** Different orderings of certain specified objects.

**Counting Rule:** If one thing can be done in $n_1$ ways and a second thing can be done in $n_2$ ways, then both things can be done together in $n_1 n_2$ ways.

**Factorial Notation:** $n! = n \cdot (n-1) \cdot (n-2) \cdot \ldots \cdot 2 \cdot 1$. The notation $n!$ is read "$n$ factorial."

**Permute:** To change position, rearrange, or reorder.

**Permutations of $n$ Things Taken $r$ at a Time:** The number of ways in which a group of $r$ objects may be chosen from a larger group of $n$ objects when the order of appearance of the $r$ objects is important.

$$P(n,r) = \frac{n!}{(n-r)!}$$

**Combinations of $n$ Things Taken $r$ at a Time:** The number of ways in which a group of $r$ objects may be chosen from a larger group of $n$ objects without regard to their order or relationship to one another within the smaller group.

$$C(n,r) = \frac{n!}{r!(n-r)!}$$

chapter 14

# APPLICATIONS OF COUNTING RULES

In Chapter 13 we considered various methods of counting, namely the counting rule, permutations, and combinations. In order to fix these ideas more clearly, we shall apply these rules and methods to various everyday problems. Notice that we have, in general, three categories of problems. In one we are considering that one thing can be done $n_1$ ways and a second thing in $n_2$ ways, in which case both things can be done together in $n_1 n_2$ ways. This method yields the counting rule. In the second category we are trying to find the number of ways that a small group of things can be chosen from a large group of things when the order of appearance in the small group affects the outcome. This type is a permutations problem. In the third category we are trying to find the number of ways that a small group of things can be chosen from a large group of things when the order of appearance in the small group does not affect the outcome. This final category encompasses combinations problems.

This chapter contains no new ideas; it merely presents problems and questions which involve all three methods of counting. We must decide which method or combination of methods applies in each case. Thus, in this chapter, problems are presented which more closely approximate real-life situations which require using the methods of Chapter 13.

### APPLICATIONS OF COUNTING RULES

_____  1. If all questions are answered in a true-false quiz of ten questions, in how many different ways may the quiz be answered?

_____  2. If three dice are tossed, how many outcomes are possible?

_____  3. If one die and two coins are tossed, how many outcomes are possible?

_____  4. How many numbers of three digits each may be formed using the numbers 1, 2, 3, 4, and 5, if repetition of digits is allowed?

_____  5. How many numbers of at least three digits (allowing no repetition of digits) may be formed using the numbers 1, 2, 3, 4, and 5?

_____  6. How many numbers of three different digits each may be formed using the numbers 1, 2, 3, 4, and 5?

_____  7. How many numbers of at most three digits (allowing no repetition of digits) may be formed using the numbers 1, 2, 3, 4, and 5?

_____  8. How many numbers of at most three digits may be formed (allowing repetition) using the numbers 1, 2, 3, 4, and 5?

_____  9. In how many ways can a club elect a president, vice-president, secretary, and treasurer from its 20 members?

_____  10. In how many ways can a club choose a committee of 5 members from its membership of 20?

_____  11. In how many ways can a club choose a committee chairman and four other committee members from its membership of 20?

_____  12. A tennis club consists of 10 boys and 8 girls. How many different mixed doubles teams are possible?

_____  13. A baseball stadium has 6 entrance gates and 8 exits. In how many ways may two men enter together and leave by separate exits?

_____  14. In how many ways can 4 blue beads, 5 red beads, and 8 green beads be arranged in a string of beads?

_____  15. In how many ways can three boys and three girls be seated in a row if they must alternate (boy-girl-boy-girl)?

_____  16. In how many ways can three boys and three girls be seated in a row?

_____  17. How many license plates can be made using

121

## APPLICATIONS OF COUNTING RULES

any three letters in the alphabet for the first three places and any two digits except zero for the last two places?

_____ 18. How many permutations can be made of the letters in the word "DUMP"?

_____ 19. How many permutations can be made of the letters of the word "LETTERS"?

_____ 20. How many permutations can be made of the letters in the word "TENNESSEE"?

_____ 21. How many permutations can be made of the letters in the word "MISSISSIPPI"?

_____ 22. How many ways are there to answer all of the questions in a 20-question multiple-choice test for which each question has 4 possible choices?

_____ 23. In how many ways can a bridge hand of 13 cards be dealt from a standard deck of 52 cards?

_____ 24. How many football games are played if each of the 7 teams plays each of the other teams in the conference exactly once?

_____ 25. From a standard deck of 52 cards, in how many ways can one obtain a bridge hand of 13 cards which will contain only aces or face cards? (There are 1 ace and 4 face cards per suit, giving a total of 20 per deck.)

_____ 26. From 6 red balls, 4 white balls, and 3 blue balls, in how many ways may one choose 2 red balls, 3 white balls, and 1 blue ball?

*chapter 15*

# INDEPENDENT EXPERIMENTS

When two coins are tossed, the outcome observed for one does not affect the outcome of the other. Likewise, independent outcomes are observed when two or three or any number of dice are tossed. And the outcomes of a coin tossed two or three or any number of times are also unaffected by previous or future occurrences—after all, the coin has no memory! **Independent experiments** are two or more experiments, the result of each of which has no effect upon the result of any of the others. There are many examples of independent experiments: a person's shoe size and the price of milkshakes, a person's grades in a mathematics course and his house number, whether or not it rains today and the Dow Jones stock average, to mention a few. There are also many examples of events or experiments that are dependent: a person's grades in mathematics and the number of hours he spent studying, a person's shoe size and his height, whether or not it rains today and whether or not it rained yesterday (because weather patterns tend to persist).

*Exercise 15-1:* In Exercise 13-7 you outlined by a tree diagram all possible outcomes that could occur when a red die and a green die are tossed. Now list all of these sample points in the form of a graph on Figure 15-1. The point "2 on the Red Die and 3 on the Green Die" has been placed for you as an example.

## INDEPENDENT EXPERIMENTS

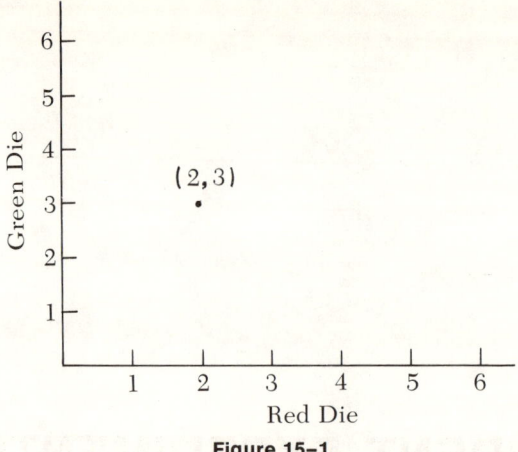

**Figure 15-1**

*Exercise 15-2:* You have already outlined in the form of a tree diagram in Exercise 13-8 the possible outcomes when a coin and a die are tossed. Now list these possible simple events as points on the graph of Figure 15-2. Label the sample points. The event "2 on the die and tails on the coin," corresponding to the point (2,T), has been placed and labeled for you as an example.

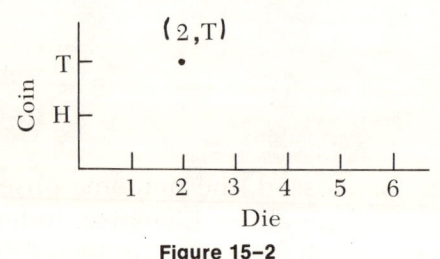

**Figure 15-2**

*Exercise 15-3:* Name three more examples of independent experiments.

*Exercise 15-4:* Would the experiments outlined in Exercises 12-1 and 12-2 (page 104) be independent? Exercise 12-1: _____; Exercise 12-2: _____. When a contractor builds a model house, is the number of bathrooms included independent of the number of bedrooms? _____
  yes or no

124

## INDEPENDENT EXPERIMENTS

The next two exercises give further examples of independent experiments. They also serve to illustrate that, when performing experiments, we often find our results to be different from what we would expect.

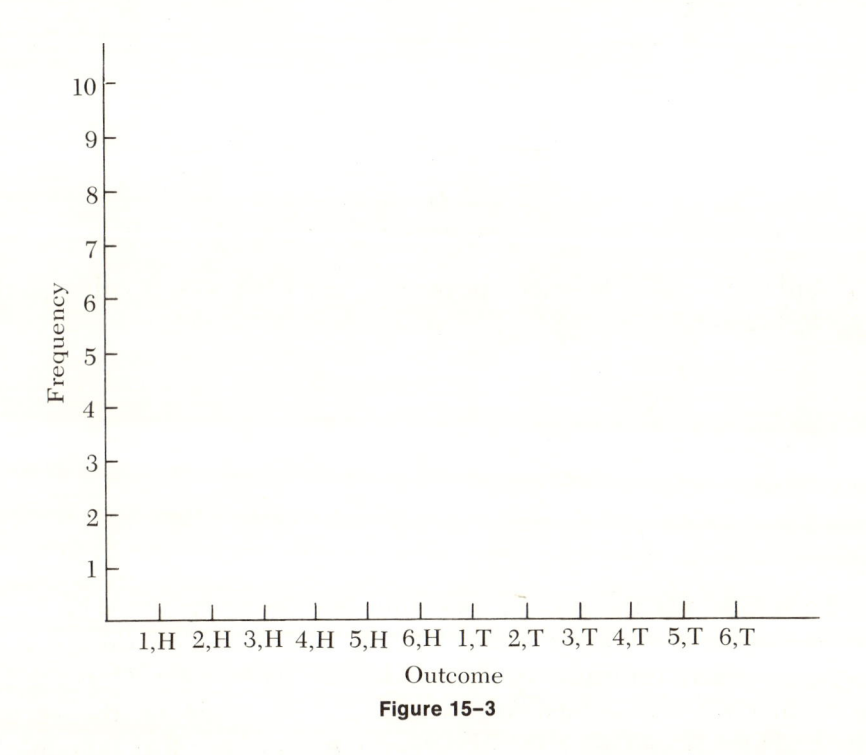

Figure 15-3

***Exercise 15-5:*** Toss a coin and a die together 60 times and record the results on the graph of Figure 15-3. Notice that there are twelve possible outcomes—$2 \cdot 6$ by the counting formula. If the coin and the die are both fair, these twelve events should be equally likely, since the outcome for the die is independent of that for the coin. By means of a dotted line, indicate what your expected result would be and comment on the similarities or differences between your observed results and the expected results.

***Exercise 15-6:*** Toss a dime and a nickel together 60 times and record the result on the graph of Figure 15-4. Notice that there are four possible outcomes—$2 \cdot 2$ by the counting formula. If both coins are fair, these four events should be equally likely, since the outcome for the dime is independent of that for the nickel. By means of a dotted line, indicate what your expected result would be and comment on

## INDEPENDENT EXPERIMENTS

**Figure 15-4**

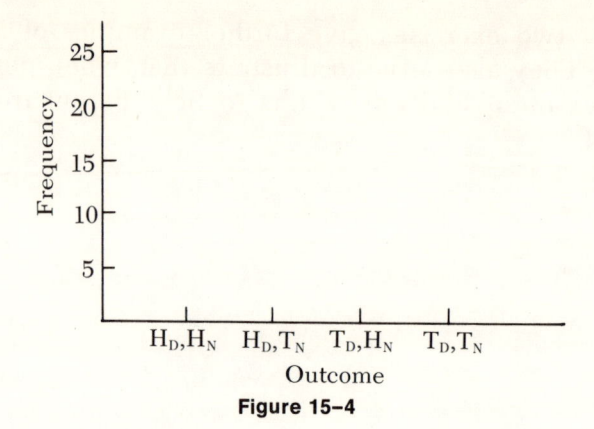

(Frequency vs Outcome: $H_D,H_N$   $H_D,T_N$   $T_D,H_N$   $T_D,T_N$)

the similarities or differences between your observed result and the expected result.

To test a missile system, two missiles might be launched, one after the other at separate times. These are not independent experiments, since any information obtained from the first launch is used to assure that the same mistakes would not be made on the second launch. Thus, because of the observed results of the first launch, the outcome of the second launch is different from what it would have been given no prior information. In other words, the results of the second launch are affected by those of the first. Whether or not a student goes out on a date a second time depends on the observed results of the first date. Whether a man buys a Ford depends on whether he was pleased with the performance of the last Ford he owned. Whether or not a student registers for English 102 next term depends on whether or not he passes English 101 this term. Whether or not a farmer plants corn in the field by the creek depends on whether the creek flooded his corn field the last time he planted corn there. These are all examples of **dependent experiments.**

Most professional gambling games are based upon combinations of independent experiments, although the card game of Black Jack is a well-known exception. Often they are based on models which have many other applications as well as the gambling application. The next two exercises suggest gambling games based upon the models developed in Exercises 15-1 and 15-2.

***Exercise 15-7:*** Suppose you are playing a game in which you toss one die, record the results, and then toss a second die. If the number on the second die is larger than that on the first die, then you win an amount of money equal to the difference. Thus if you toss (3,5) you win $2. However, if the number on the second die is less than or equal to that on the first, you win $0 and lose what you paid to play the game. Make a table of all possible wins and decide whether it would be profitable to pay $2 per toss of two dice to play the game. Three possible values have been filled in for you as examples. Of course, since the dice are fair, each of the 36 possible events is equally likely.

TABLE OF WINS

Second Die

|   | 1 | 2 | 3 | 4 | 5 | 6 |
|---|---|---|---|---|---|---|
| 6 | 5 |   |   |   |   |   |
| 5 |   |   | 2 |   |   |   |
| 4 |   |   |   |   |   |   |
| 3 |   |   |   |   |   |   |
| 2 |   |   |   |   |   |   |
| 1 |   |   |   |   |   | 0 |

First Die

**Exercise 15–8:** Suppose you are playing a game which costs $5 to play. The game consists of first tossing a coin. If the coin lands T, you lose, but if the coin lands H, you then toss a die and win twice the number on the upturned face in dollars. Thus (H,1) means you pay $5, and win $2; (H,6) means you pay $5, and win $12; (T,3) means you pay $5 and win nothing. Make a table of all possible wins or losses and decide whether this would be a profitable game to play. Since the coin and die are fair, each of the twelve possible sample points is equally likely. Three possible values have been filled in for you as examples.

TABLE OF WINS

| Coin |   | 1 | 2 | 3 | 4 | 5 | 6 |
|------|---|---|---|---|---|---|---|
|      | H | 2 |   |   |   |   | 12 |
|      | T |   |   | 0 |   |   |   |

Die

The probabilities of events from independent experiments are usually rather easy to calculate. Exercises 15–9 and 15–10 illustrate another kind of model often used in gambling situations.

## INDEPENDENT EXPERIMENTS

**Exercise 15-9:** Which would be more likely, getting a double [(1,1), (2,2), (3,3), (4,4), (5,5), or (6,6)] when two dice are tossed; or getting a triple [such as (1,1,1), (2,2,2), (3,3,3), (4,4,4), (5,5,5), or (6,6,6)] when three dice are tossed?

**Exercise 15-10:** Which is more likely, getting a double [such as (1,1), (2,2), (3,3), (4,4), (5,5), or (6,6)] when you toss two dice; or getting a double [such as (H,H) or (T,T)] when you toss two coins?

**Exercise 15-11:** Suppose a fair coin is tossed 10 times and 10 heads have appeared. What are the chances a tail will appear on the 11th toss? _____ Why? _____

**Exercise 15-12:** Suppose a fair die is tossed 3 times and a "2" has appeared each time. What are the chances that on the fourth toss a "2" will appear? _____ Why? _____

_____

**Exercise 15-13:** Toss a coin 100 times and record the results in a sequence, such as HTHHHTT. Underline the longest run of heads and the longest run of tails. Is it common or uncommon to get runs of, say, 5 or 6 H's or 5 or 6 T's in a row? _____

Is it common to have alternating sequences like HTHTHT? _____

# INDEPENDENT EXPERIMENTS

**Exercise 15-14:** Toss a die 100 times and record the results in a sequence, like 1124163. Underline the longest run of even numbers and the longest run of odd numbers. Are long runs common or uncommon? _____ Is it more or less common to have a long run of even numbers than to have a long run of any single number?
_____

Why? _____
_____

**TERMS TO REMEMBER**

**Independent Experiments:** Two or more experiments, the outcome of each of which has no effect upon the outcome of any other.

**Dependent Experiments:** Two or more experiments that are not independent.

129

chapter 16

# THE MANN-WHITNEY *U* TEST

When we applied the sign test in Chapter 11 or the signed ranks test in Chapter 12 to determine whether observations taken from two groups exhibited different behavior in terms of location measures, one of our basic assumptions was that each member of one group was matched or paired with a specific member of the other group. In other words we expected the treatment effects, or responses to treatment, of these matched individuals to be similar. We expected this conformity partly because each individual's response depended to a degree upon some factor which also affected the response of the other member of the matched pair. In the language of Chapter 15, the groups were dependent.

Suppose that we have two groups from which we would expect independent measures. Then we would expect the response of any subject of one group to be unaffected by the responses of any of the subjects of the other group. In this case the responses would depend solely upon the treatment applied, and we would expect no relationship to exist between individuals of different groups. In fact, the numbers of observations taken for the two groups may even be different, whereas for the matched pairs, there obviously had to be the same number of subjects in both groups. A few examples might illustrate the differences between dependent and independent measures taken on two groups.

EXAMPLE 16–1. Suppose that an experimenter compares the weights of pigs at the age of 6 weeks with the weights of the same pigs after 5 months of a certain diet ration. These measures would be dependent or paired. Each pig's final weight may well depend on its initial weight.

EXAMPLE 16–2. Next consider comparing the weights of pigs from one litter that have been given diet A with the weights of the pigs from a second litter that have been given diet B. These measures would be independent since the pigs came from two different litters.

EXAMPLE 16–3. Suppose that ten sets of identical twins are separated into two classes, each containing only one of each pair of twins. One class is taught by Method A and one by Method B. The observations are achievement scores at the end of the term. These scores would be dependent because identical twins would have approximately the same IQ and learning abilities.

EXAMPLE 16–4. Suppose that December sales totals of retail department stores in Kentucky are compared with December sales totals of retail department stores in Florida for a particular year. These observations would be independent.

If the measurement scale is at least ordinal, one may use the Mann-Whitney $U$ Test to determine whether two independent groups have responded differently to treatments. Thus the Mann-Whitney $U$ Test is similar to the signed ranks test in that they both assume the same measurement scale. The two tests differ only in that the signed ranks test pertains to two matched groups, while the Mann-Whitney $U$ Test is used when the two samples are independent. Both determine whether the measures of location of the observations for the two groups are different. In Chapter 25 we shall consider a test which can be used for two independent samples having a nominal measurement scale.

The procedure to follow for the Mann-Whitney $U$ Test is quite simple.

1. List the observations for both samples together in order of increasing rank or size. If there are negative measurements, they are considered to be smaller than any of the positive measurements and would be listed first. In some manner, record which observation comes from which sample. One might underline all the observations of one sample and leave those of the other sample not underlined, or else one might label or subscript each observation to identify it.
2. Arbitrarily assign the number "1" to one sample and "2" to the other. Then record $n_1 =$ the size of sample 1 and $n_2 =$ the size of sample 2.
3. Assign ranks to the observations according to their size, again identifying which ranks come from which sample. If some observations are tied, assign the average of the tied ranks to all ob-

# THE MANN-WHITNEY U TEST

servations that have the same size, as was done for the signed ranks test.

4. Calculate $R_1$ = the sum of the ranks of sample 1, and $R_2$ = the sum of the ranks of sample 2.
5. Calculate $U_1 = n_1 n_2 + \dfrac{n_1(n_1+1)}{2} - R_1$ and $U_2 = n_1 n_2 + \dfrac{n_2(n_2+1)}{2} - R_2$. As a check, it will always be true that $U_1 = n_1 n_2 - U_2$.
6. Let $U$ equal the smaller of the two values $U_1$ and $U_2$.
7. If $U$ is not larger than the critical value given in Table V of the appendix, then one may conclude that a calculated $U$-value this small would occur by chance less than ten per cent of the time if in fact the measure of location of one group did not differ from that of the other. In other words, either the experimental observations have yielded a result which would occur less than ten per cent of the time if there were in fact no true differences, or else there do exist true differences between the two responses to treatment. If $U$ is larger than this critical value, then any difference observed could be due to random sampling error and would not be considered a true difference.

EXAMPLE 16-5. Suppose that students, assigned randomly to two small high school chemistry classes, were taught by two different methods. In Class "Demo," the instructor demonstrated the experimental techniques, while in Class "Exp," the students themselves performed the experiments. Following these two methods of instruction, each class took an achievement test designed to measure knowledge of the basic principles of chemistry. Their letter grades on the test were as shown in the following table. Did the two groups respond the same to the two teaching techniques? In other words, from the given data, could one conclude that students learn the basic principles of chemistry equally well when taught by a demonstration method as when taught by an experimentation method?

| Demo | C, | B, | C+, | D, | A−, | E, | C−, | B−, | C, | D− |
|---|---|---|---|---|---|---|---|---|---|---|
| Exp | C−, | B, | B+, | A, | D, | A−, | B, | C, | A+ | |

*Step 1:* (Grades from Exp are underlined) E, D−, <u>D</u>, D, C−, <u>C−</u>, <u>C</u>, C, C, C+, B−, B, <u>B</u>, <u>B</u>, <u>B+</u>, A−, <u>A−</u>, <u>A</u>, <u>A+</u>.

*Step 2:* $n_1 = n_{\text{Demo}} = 10$; $n_2 = n_{\text{Exp}} = 9$

*Step 3:* (Ranks from Exp are underlined) 1, 2, 3.5, <u>3.5</u>, 5.5, <u>5.5</u>, <u>8</u>, 8, 8, 10, 11, 13, <u>13</u>, <u>13</u>, <u>15</u>, 16.5, <u>16.5</u>, <u>18</u>, <u>19</u>.

*Step 4:* $R_1 = 78.5$; $R_2 = 111.5$.

*Step 5:* $\quad U_1 = (10)(9) + \dfrac{(10)(11)}{2} - 78.5 = 66.5$

$U_2 = (10)(9) + \dfrac{(9)(10)}{2} - 111.5 = 23.5$

Check: $\quad U_1 = (10)(9) - 23.5 = 66.5$

*Step 6:* $\quad U = 23.5$

*Step 7:* 23.5 < 24 from Table V. Thus we may conclude that the sample evidence implies that either we have observed an extreme event or the group taught by the experimental approach learned significantly more basic principles of chemistry.

The next two exercises provide experience in performing the Mann-Whitney $U$ Test. To "cut" a deck of cards means to put the top part of the deck—about half of the cards—on the bottom. Notice that cutting cards does not tend to mix the cards very well.

**Exercise 16-1:** Take two standard bridge decks of cards. Shuffle one deck thoroughly and cut it once. Then lay the deck face down. Turn up the first ten cards from the top of the deck and if the card is not a club, record as a score the number on the card. Face cards count "10" and the aces count "one." If the card is a club, the score is zero for that card. Thus, for example, if the first ten cards are 3H (three of hearts), 4S (4 of spades), JS (jack of spades), QD (queen of diamonds), 1C, 4H, 10D, KH, 3D, 8C, the scores would be 3, 4, 10, 10, 0, 4, 10, 10, 3, 0. Remove all the clubs from the second deck. This will be the "treated" deck. Place the deck face down and then put the clubs face down on the top of the deck. Do not shuffle, but cut the cards four times. Turn up the first 10 cards and record the scores as before. Use a Mann-Whitney $U$ Test to see whether the two groups yield different scores. Shuffle well—at least ten times—and repeat the entire experiment, cutting the cards of the "treated" deck three times instead of four. Are the results the same? Why or why not?

# THE MANN-WHITNEY U TEST

***Exercise 16-2:*** Take two standard bridge decks of cards. Shuffle one deck thoroughly, cut it once, and lay the deck face down. Turn up the first ten cards from the top of the deck and record as a score the number on the card unless the card is a heart. Assume that the face cards count "10" and the aces count "one." If the card is a heart, the score is zero for that card. Thus, for example, if the first ten cards are 3H (3 of hearts), 4S (4 of spades), JS (jack of spades), QD (queen of diamonds), 1C, 4H, 10D, KH, 3D, 8C, the scores would be 0, 4, 10, 10, 1, 0, 10, 0, 3, 8. From the second deck, the "treated deck," remove all the hearts. Place the deck face down and then put the hearts face down on top of the deck. Do not shuffle, but cut the cards four times. Turn up the first 10 cards and record the scores as before. Use a Mann-Whitney $U$ Test to see whether the two groups yield different scores. Shuffle the cards thoroughly—at least ten times—and repeat the experiment, cutting the cards of the "treated" deck three times instead of four. Are the results the same? Why or why not?

***Exercise 16-3:*** Gather the following information from each person of a group of about twenty people, being sure that the group includes both sexes:
1. M or F, indicating the sex of the person.
2. A number, in hours, indicating approximately how much time that person spends each week watching TV.

Let "M" be group 1 and "F" be group 2. Use a Mann-Whitney $U$ Test to determine whether men and women spend the same amount of time per week watching TV.

Do you believe that most people can accurately remember the number of hours per week they spend watching TV? _____

Do you believe your sample of people is truly representative of the entire population of the United States? _____

If not, why not? _____

_____

What group does your sample represent? _____

_____

***Exercise 16-4:*** Gather the following information from each person of a group of about twenty people, being sure that the group includes people over 30 and people under 30:
1. Indication of age—either "over 30" or "under 30."
2. A number, in hours, indicating approximately how much time that person spends each week watching TV.

Label the "over 30" group 1 and the "under 30" group 2. Use a Mann-Whitney *U* Test to determine whether people of different age groups spend the same amount of time per week watching TV.

### THE MANN-WHITNEY U TEST

Do you believe that most people can accurately remember the number of hours per week they spend watching TV? _____

Do you believe your sample of people is truly representative of the entire population of the United States? _____

If not, why not? _____

_____

What group does your sample represent? _____

_____

chapter 17

# MEASURES OF DISPERSION

In this chapter we will study measures of the dispersion or spread among a group of observations obtained from sample data. First, though, we need to learn what we mean by "dispersion" and why we are interested in measuring it.

EXAMPLE 17-1. Suppose a buyer for a manufacturing concern had to choose among three suppliers of bolts. If all the suppliers' bolts averaged 1/2 inch in diameter, then the criterion for a choice would be to find which supplier's bolts showed the least variation in diameter. To do so, he could randomly choose 5 bolts each from boxes labeled "1/2 inch bolts" made by suppliers A, B, and C, and carefully measure the diameter of the bolts. Suppose the measurements, in inches, were as follows:

A: 0.52, 0.50, 0.56, 0.43, 0.49

B: 0.50, 0.53, 0.50, 0.49, 0.48

C: 0.49, 0.50, 0.50, 0.50, 0.51

Which supplier should he choose? (Answer: C, because the measurements vary less.)

EXAMPLE 17-2. Suppose a personnel manager is interviewing applicants for the position of cashier in a department store. He gives

## MEASURES OF DISPERSION

each of four applicants 5 columns of numbers, in dollars and cents, to add. The columns all really have the same amounts in different order, so that all totals should be $11.00, if correct. The totals obtained by the 4 applicants are:

A: 10.99, 11.03, 11.00, 11.02, 10.96

B: 11.00, 11.00, 11.01, 10.99, 11.00

C: 10.95, 10.90, 11.15, 11.02, 10.98

D: 11.10, 11.05, 11.07, 10.95, 11.00

Which applicant should he hire? (Answer: B, because his totals vary less; D does not even have an average value of $11.00.)

There exist many similar examples. In each case there are differences among the numbers—but the smaller the differences, the more precise the measurement, or the better the quality control, or the more accurate the individual, and so on. In experimental data, one wishes to control error as much as possible. To obtain a measure of his precision, the experimenter may calculate the measure of dispersion that is most appropriate for his data. He will want either the range, which is the simplest measure; the average deviation, which is used with the median or with outliers; or the standard deviation, which is the most often used. A simple measure of central tendency is meaningless unless accompanied by a measure indicating the degree of precision with which the observations were taken. That degree of precision is specified by a measure of dispersion.

Consider, for example, the observations 6, 4, 8, 2, 4, 12, which when put in ascending order are 2, 4, 4, 6, 8, 12. Call these "Sample A" and compare them with two other samples of observations as shown in Table 17–1. All of these samples have the same mean; thus they do not differ in central tendency. Clearly, however, they do differ considerably in dispersion. Notice that the range of the numbers is different: for A, the range is $12 - 2 = 10$; for B, it is only $6.3 - 5.7 = 0.6$; and for C, it is $12 - 0 = 12$. The **range** is the distance on a number scale between the largest and the smallest observations. Thus the range equals $(X_L - X_S)$. If the midrange has been used as a measure

**TABLE 17–1** THREE SETS OF SAMPLE OBSERVATIONS WHICH HAVE THE SAME MEAN.

| Sample | Observations | | | | | | $\bar{x} = \frac{\Sigma X}{n}$ |
|---|---|---|---|---|---|---|---|
| A | 2 | 4 | 4 | 6 | 8 | 12 | 6 |
| B | 5.7 | 5.9 | 6.0 | 6.0 | 6.1 | 6.3 | 6 |
| C | 0 | 0 | 0 | 12 | 12 | 12 | 6 |

## MEASURES OF DISPERSION

of location, then the range is usually employed as the corresponding measure of dispersion. The range has the advantage that it is easy to calculate, but it has the disadvantage that it is strongly affected by outliers.

In order to illustrate graphically the meaning of a measure of dispersion, we may draw a number line and mark it off in the units 0 through 12. Next we may fill in the points from Table 17-1 as shown in Figure 17-1.

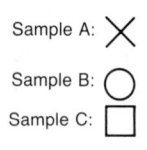

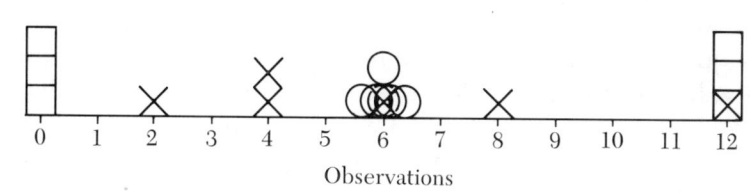

Sample A: ✕
Sample B: ○
Sample C: ☐

**Figure 17-1** Illustration of the dispersions of samples A, B, and C from Table 17-1.

For Sample A, the circles representing its data observations are spread out from the point of central tendency, but not nearly so much as are the squares of Sample C. Sample B, on the other hand, is closely packed around its point of central tendency. All measures of dispersion should indicate C as having the greatest, A the next, and B the least dispersion or spread. As we have already seen, the range does give us this very description.

Before pursuing the topic of dispersion any further, we need to practice working with summations. Consider again the data from Sample A of Table 17-1, which we may denote as follows:

$$X_1 = 6, \ X_2 = 4, \ X_3 = 8, \ X_4 = 2, \ X_5 = 4, \ X_6 = 12.$$

To indicate that we want to add all six values, we could write $X_1 + X_2 + X_3 + X_4 + X_5 + X_6$; but a much simpler method is to write $\Sigma X$. The symbol $\Sigma$ is a kind of mathematical shorthand notation for "take the sum of." The expression $\Sigma X^2$ would mean "take the sum of the

## MEASURES OF DISPERSION

squares of all of the $X$'s." Thus, for Sample A, $\Sigma X = 6+4+8+2+4+12 = 36$; the arithmetic mean (which we give the symbol $\bar{x}$) is given in this notation as

$$\bar{x} = \frac{\Sigma X}{n} = \frac{36}{6} = 6.$$

We use parentheses to show which operation should be performed first when more than one operation must be done. Thus, while $\Sigma X^2 = 36+16+64+4+16+144 = 280$, the notation $(\Sigma X)^2$ means $36^2 = 1296$. Notice that $(\Sigma X)^2$ says to square the sum, while $\Sigma X^2$ says to sum the squares. They do *not* yield the same results! Also, notice that $\Sigma(X+2) = 8+6+10+4+6+14 = 48$, while $\Sigma X + 2 = 36+2 = 38$; and $\Sigma(X-6)^2 = (0)^2 + (-2)^2 + (2)^2 + (-4)^2 + (-2)^2 + (6)^2 = 64$.

Measures of dispersion often involve another basic mathematical tool—the absolute value. The **absolute value** of a quantity is its size regardless of sign, and is denoted by placing vertical lines to either side of the quantity. For example, $|-6| = 6$ and $|2| = 2$, while $|a| = a$ if $a$ is positive and $|a| = -a$ if $a$ is a negative quantity.

Table VI of the appendix is a table of squares and square roots, which we shall also need to calculate measures of dispersion. To obtain the square root of a number, look for that number in the first column. If it is in the first column, read $\sqrt{N}$ in the third column of the same row as $N$. If one tenth of the number is in the first column, read the root in the fourth column of that row. If the root desired is still not obtainable, move its decimal in groups of 2 until either the desired value or one tenth of it appears in the first column. Then read the appropriate root from Table VI and move its decimal one place for every two places that the decimal had to be moved in order to obtain a value $N$ in the first column.

EXAMPLE 17-3. Using the numbers 1, 3, 9, 7, 5, verify (a) through (h) and using Table VI, verify (i) through (n):

(a) $\Sigma X = 25$          (f) The mean $= 5$          (k) $\sqrt{.0046} = .068$
(b) $\Sigma X^2 = 165$       (g) The median $= 5$        (l) $\sqrt{23{,}150.} = 152.$
(c) $\Sigma(X+2) = 35$       (h) The mode $=$ none       (m) $\sqrt{.213} = .146$
(d) $\Sigma(X-5)^2 = 40$     (i) $\sqrt{1.43} = 1.20$    (n) $\sqrt{316.4} = 17.8$
(e) $\Sigma|X-5| = 12$       (j) $\sqrt{27.4} = 5.23$

***Exercise 17-1:*** Find the values of the following quantities.

(a) $|-10| = $ _____

(b) $|6| = $ _____

(c) $|3| - |7| = $ _____

## MEASURES OF DISPERSION

(d) $|3-7| = $ _____

(e) $\sqrt{6.73} = $ _____

(f) $\sqrt{256} = $ _____

**Exercise 17-2:** Find the values of the following quantities.

(a) $|-8| = $ _____

(b) $|5| = $ _____

(c) $|2| - |6| = $ _____

(d) $|2-6| = $ _____

(e) $\sqrt{4.82} = $ _____

(f) $\sqrt{397} = $ _____

**Exercise 17-3:** Given the sample observations $X_1 = 1$, $X_2 = 7$, $X_3 = 2$, $X_4 = 3$, $X_5 = 7$, find:

(a) $\Sigma X$ = _____

(b) $\Sigma(X-4)$ = _____

(c) $\Sigma(X+1)$ = _____

(d) $\Sigma(X^2+1)$ = _____

(e) $\Sigma X^2$ = _____

(f) $\Sigma(X-4)^2$ = _____

(g) $\Sigma \dfrac{(X-4)^2}{4}$ = _____

(h) the mean = _____

(i) the median = _____

(j) the mode = _____

(k) $\Sigma|X-4|$ = _____

**Exercise 17-4:** Given the sample observations $X_1 = 2$, $X_2 = 8$, $X_3 = 4$, $X_4 = 9$, $X_5 = 2$, find:

(a) $\Sigma X$ = _____

(b) $\Sigma(X-5)$ = _____

(c) $\Sigma(X-1)$ = _____

(d) $\Sigma(X^2+1)$ = _____

(e) $\Sigma X^2$ = _____

(f) $\Sigma(X-5)^2$ = _____

(g) $\Sigma \dfrac{(X-5)^2}{4}$ = _____

(h) the mean = _____

(i) the median = _____

(j) the mode = _____

(k) $\Sigma|X-5|$ = _____

# MEASURES OF DISPERSION

Now that we are prepared for the arithmetic required, we may continue our investigation of sample variation. We saw that the range, which was discussed earlier in this chapter, is easily biased by outliers. A measure of dispersion that is not so adversely affected by outliers is called the **average deviation**, or **mean deviation**. To find the average deviation of a group of observations:

(1) Find the median of the observations.
(2) Calculate the difference between each observation and the median. These differences are called **deviations from the median** and are denoted by $\tilde{d}$, just as the observations are denoted by $X$. Thus for any observation $X$, we define $\tilde{d} = X -$ median.
(3) Add all of the *absolute values* of the deviations and divide this sum by the number of observations, denoted by $n$. The result is the average deviation; that is, average deviation = $\frac{\Sigma |\tilde{d}|}{n} = \frac{\Sigma |X - \text{median}|}{n}$.

Whenever the median is used as a measure of central tendency, the average deviation is normally used as the corresponding measure of dispersion. Consequently, the average deviation is used with ordinal data or with interval data that have outliers when a descriptive measure is desired.

EXAMPLE 17-4. As an example of the calculation of the average deviation, consider Sample A of Table 17-1. Its median is 5.

| X (Observations) | $\tilde{d} = X -$ median (Deviations) | $\|\tilde{d}\|$ (Absolute Values of Deviations) |
|---|---|---|
| 2 | −3 | 3 |
| 4 | −1 | 1 |
| 4 | −1 | 1 |
| 6 | 1 | 1 |
| 8 | 3 | 3 |
| 12 | 7 | 7 |
| $\bar{x} = \frac{\Sigma X}{n} = 6$ | | $\Sigma\|\tilde{d}\| = 16$ |

The average deviation is $\frac{\Sigma |\tilde{d}|}{n} = \frac{16}{6} = 2.7$.

**Exercise 17-5:** Calculate the average deviation for Sample B of Table 17-1. Sample B: $\frac{\Sigma |\tilde{d}|}{n} =$ _____

# MEASURES OF DISPERSION

**Exercise 17-6:** Calculate the average deviation for Sample C, Table 17-1. Sample C: $\frac{\Sigma|d|}{n} = $ _____



The primary disadvantage of the average deviation is that absolute values are difficult to work with. For this reason, the most often used measure of dispersion is the standard deviation. This is calculated as follows:
(1) Find the mean of the observations.
(2) Calculate the difference between each observation and the mean. These differences are called **deviations from the mean** and are denoted by $d$. Thus, for any observation $X$, we define $d = X - \bar{x}$.
(3) Add all the squares of these deviations and divide the sum by one less than the total number of observations (that is, by $n-1$). The result is the **variance**.
(4) Take the square root of the variance to obtain the **standard deviation**. Thus, the standard deviation, which we denote by $s$, is

$$s = \sqrt{(\Sigma d^2)/(n-1)},$$

and the variance is $s^2 = (\Sigma d^2)/(n-1)$.

Whenever the mean is the appropriate measure of central tendency, one should use the standard deviation as the corresponding measure of dispersion.

EXAMPLE 17-5. In Sample A of Table 17-1, the mean is 6, and thus we can calculate the standard deviation as follows:

| X (Observations) | $d = X - \bar{x}$ (Deviations) | $d^2$ (Squared Deviations) |
|---|---|---|
| 2 | −4 | 16 |
| 4 | −2 | 4 |
| 4 | −2 | 4 |
| 6 | 0 | 0 |
| 8 | 2 | 4 |
| 12 | 6 | 36 |
| $\bar{x} = \frac{\Sigma X}{n} = 6$ | $\Sigma d = 0$ | $\Sigma d^2 = 64$ |

The standard deviation is $s = \sqrt{\frac{16+4+4+0+4+36}{5}} = \sqrt{\frac{64}{5}} = \sqrt{12.8} = 3.6$, having obtained the square root from Table VI.

# MEASURES OF DISPERSION

**Exercise 17-7:** Calculate the standard deviation for Sample B of Table 17-1.

$s_B =$ _____

**Exercise 17-8:** Calculate the standard deviation for Sample C of Table 17-1.

$s_C =$ _____

**Exercise 17-9:** Calculate the mean, median, range, average deviation, and standard deviation separately for each of the two groups of observations as given by Exercise 16-1, page 000, and fill in the following table:

*Data Gathered in Exercise 16-1*

| Group | MEAN | MEDIAN | RANGE | AVERAGE DEVIATION | STANDARD DEVIATION |
|---|---|---|---|---|---|
| 1 | | | | | |
| 2 | | | | | |

## MEASURES OF DISPERSION

Now graph the data for the two groups using two superimposed polygons. Identify the separate groups. You may wish to use a solid line for one and a dashed line for the other; or they could be done with different colored lines.

***Exercise 17-10:*** Calculate the mean, median, range, average deviation, and standard deviation separately for each of the two groups of observations as given by Exercise 16-2, page 134, and fill in the following table:

| | Data Gathered in Exercise 16-2 | | | | |
|---|---|---|---|---|---|
| Group | MEAN | MEDIAN | RANGE | AVERAGE DEVIATION | STANDARD DEVIATION |
| 1 | | | | | |
| 2 | | | | | |

Now graph the data for the two groups using two superimposed polygons. Identify the separate groups. You may wish to use a solid line for one and a dashed line for the other; or they could be done with different colored lines.

## MEASURES OF DISPERSION

***Exercise 17-11:*** Calculate the mean, median, range, average deviation and standard deviation separately for each of the two groups for which observations were taken in Exercise 16-3, page 134, and fill in the following table:

Data Gathered in Exercise 16-3

| Group | MEAN | MEDIAN | RANGE | AVERAGE DEVIATION | STANDARD DEVIATION |
|---|---|---|---|---|---|
| 1 | | | | | |
| 2 | | | | | |

***Exercise 17-12:*** Calculate the mean, median, range, average deviation, and standard deviation separately for each of the two groups for which observations were taken in Exercise 16-4, page 135, and fill in the following table:

Data Gathered in Exercise 16-4

| Group | MEAN | MEDIAN | RANGE | AVERAGE DEVIATION | STANDARD DEVIATION |
|---|---|---|---|---|---|
| 1 | | | | | |
| 2 | | | | | |

***Exercise 17-13:*** What measure of dispersion do you think is best to use for:

(a) Daily temperatures _____

(b) Heights of male athletes _____

(c) Heights of all men on a college campus _____

(d) Daily stock prices _____

(e) Make of car driven on a university campus (Does it even make sense to consider "spread" or "dispersion" in this example?)

_____

(f) Year of car driven on a university campus _____

_____

# MEASURES OF DISPERSION

**Exercise 17-14:** What measure of dispersion do you think is best to use for:

(a) Daily body temperature of a patient in a hospital _____

_____

(b) Amount of milk yield per year of a dairy cow _____

(c) Typing speed of a stenographer _____

(d) Grade point average of a student over an academic year

_____

(e) Depth of a river _____
(f) Type of sandwich ordered at a restaurant (Can "spread" or

"dispersion" even be measured in this case?) _____

_____

**TERMS TO REMEMBER**

**Range:** The difference between the largest observation ($X_L$) and the smallest observation ($X_S$) in a set of sample data, or $X_L - X_S$.

**Σ:** Notation meaning "take the sum of" or "add."

**x̄:** Notation meaning the arithmetic average; $\bar{x} = \frac{\Sigma X}{n}$, where $n$ is the total number of observations $X$.

**Absolute Value:** The magnitude of a number, ignoring its sign.

**Average Deviation:** $\frac{\Sigma |\tilde{d}|}{n}$, where the $\tilde{d}$ are the deviations from the median for all $n$ sample observations; a measure of dispersion.

**Mean Deviation:** Same as average deviation.

**Deviation from the Median:** $\tilde{d} = X -$ median for all sample observations $X$.

**Deviation from the Mean:** $d = X - \bar{x}$ for all sample observations $X$.

**Variance:** $\frac{\Sigma d^2}{n-1}$ = the square of the standard deviation.

**Standard Deviation:** $\sqrt{\frac{\Sigma d^2}{n-1}}$, where the $d$ are the deviations from the mean for all $n$ sample observations; a measure of dispersion.

chapter 18

# A TEST FOR DISPERSION

In Chapter 16 we discussed the use of the Mann-Whitney U test to find out whether the measures of central tendency of two independent groups were different. With only slight modification, we may use the same test to find out whether two independent groups of ordinal or interval measures have different measures of dispersion. To do so we must be able to assume that the measure of central tendency is the same for both groups of observations.

The procedure is as follows:

(1) Arrange the observations from both groups in order of increasing or decreasing size, identifying those which come from one of the groups just as was done in Chapter 16.

(2) Specify which is the "first" and which is the "second" sample and thereby specify $n_1 = $ the total number of observations in the first sample and $n_2 = $ the total number of observations in the second sample.

(3) Assign ranks in the following alternating manner: assign rank 1 to the smallest value; assign ranks 2 and 3 to the largest and the second-largest values; assign ranks 4 and 5 to the second-smallest and the third-smallest; and so forth. In case of ties, attempt to group the tied scores in such a way as to provide a degree of balance in the alternation scheme. In other words, try to have about as many small ranks assigned to large as to

small values. This test loses precision if there are several ties present.

(4) Calculate $R_1$ = the sum of the ranks of sample 1 and $R_2$ = the sum of the ranks of sample 2. Next, calculate

$$U_1 = n_1 n_2 + \frac{n_1(n_1 + 1)}{2} - R_1$$

and

$$U_2 = n_1 n_2 + \frac{n_2(n_2 + 1)}{2} - R_2.$$

(5) As a check, be sure that $U_2 = n_1 n_2 - U_1$.
(6) Let $U$ equal the smaller of the two values, $U_1$ and $U_2$, and use Table V of the appendix to decide whether the two samples have equal dispersions.

If the calculated value of $U$ is not larger than that given in Table V for the sample sizes $n_1$ and $n_2$, then one would suspect from the sample evidence that the two groups do have different dispersions. If $U$ is larger than the tabled value, then one would assume that the sample dispersions are the same and that any actual observed differences in variation are due to chance fluctuations.

In other words, in this chapter no new methods appear. Instead we merely use the methods presented in Chapter 16 in a different way. Suppose, for example, that two samples differ in dispersion but in no other way. Then, when the observations from the two samples are ordered together, one sample's observations would be "bunched" in the midst of the observations from the other. The most extreme case would occur when all the observations of one sample were either numerically less than the smallest observation of the other sample or else greater than the largest observation of the other sample, yielding a pattern like AAAAAABBBBBBBBBBAAAAA. When the ranks are assigned to such samples, all the A's would have the smallest ranks and all B's would have the largest ranks, resulting in the detection of an extreme difference between the sample dispersions.

EXAMPLE 18–1. Given two groups of data observations as follows, decide whether the groups have equal dispersions.

A: 32, 29, 56, 38, 43, 49, 36     $\bar{x}_A = 40.4$

B: 35, 42, 39, 45, 30, 37     $\bar{x}_B = 38.0$

Although the means differ numerically, they are near enough to the same value to be able to assume that the differences are only due to random fluctuations. This statement could be validated by testing the means for equality using the Mann-Whitney test. Let us order the data, underlining the A observations, and assign the ranks.

## A TEST FOR DISPERSION

| Ordered Data | Ranks |
|---|---|
| <u>29</u> | <u>1</u> |
| 30 | 4 |
| <u>32</u> | <u>5</u> |
| 35 | 8 |
| <u>36</u> | <u>9</u> |
| 37 | 12 |
| <u>38</u> | <u>13</u> |
| 39 | 11 |
| 42 | 10 |
| <u>43</u> | <u>7</u> |
| 45 | 6 |
| <u>49</u> | <u>3</u> |
| <u>56</u> | <u>2</u> |

$n_1 = 7$
$R_1 = 1 + 5 + 9 + 13 + 7 + 3 + 2 = 40$
$n_2 = 6$
$R_2 = 4 + 8 + 12 + 11 + 10 + 6 = 51$
$U_1 = 7 \cdot 6 + \frac{7(8)}{2} - 40 = 30$
$U_2 = 7 \cdot 6 + \frac{6(7)}{2} - 51 = 12 = 42 - 30$ (check)
$U = 12$

Critical value of $U$ from Table V is 8.
Conclusion: Based on the sample data, there is no reason to suspect unequal dispersions.

Perhaps the greatest difficulty occurs when ties are present. If three or more scores are tied for the same rank, we cannot split them. Instead we try to balance tied ranks at the higher and lower ends whenever possible, as is illustrated in Examples 18-2 and 18-3. Because of different possible ways of assigning tied ranks, one person's results in terms of a test statistic may differ slightly from those of someone else performing the same test with the same method. This lack of precision is one of the prices we pay for having so simple a test procedure. In the example and exercises that follow you may wish to assign ranks in various patterns just to see how much of a difference in the final result occurs when tied ranks are assigned in varying manners.

EXAMPLE 18-2. Given two groups of data observations as follows, test whether the groups have equal dispersions. Notice the presence of ties.

A: 1.3, 4.6, 2.2, 1.3, 3.7, 5.1   $\bar{x}_A = 3.03$
B: 1.9, 5.1, 0.2, 4.8, 6.3, 0.2, 7.6   $\bar{x}_B = 3.73$

The means are near enough to the same value for their differences to

# A TEST FOR DISPERSION

be considered random error. We could test their equality using a Mann-Whitney $U$ Test. Let us order the data, underlining the A observations, and assign ranks in two different ways.

| Ordered Data | Ranks | Alternate Assignment of Ranks (not well-balanced) |
|---|---|---|
| 0.2 | 1.5 | 2.5 |
| 0.2 | 1.5 | 2.5 |
| <u>1.3</u> | <u>5.5</u> | <u>7.5</u> |
| <u>1.3</u> | <u>5.5</u> | <u>7.5</u> |
| 1.9 | 9 | 11 |
| <u>2.2</u> | <u>10</u> | <u>12</u> |
| <u>3.7</u> | <u>13</u> | <u>13</u> |
| <u>4.6</u> | <u>12</u> | <u>10</u> |
| 4.8 | 11 | 9 |
| 5.1 | 7.5 | 5.5 |
| <u>5.1</u> | <u>7.5</u> | <u>5.5</u> |
| 6.3 | 4 | 4 |
| 7.6 | 3 | 1 |

$n_1 = 6$
$R_1 = 53.5$
$n_2 = 7$
$R_2 = 37.5$
$U_1 = 6 \cdot 7 + \dfrac{6(7)}{2} - 53.5 = 9.5$
$U_2 = 6 \cdot 7 + \dfrac{7(8)}{2} - 37.5 = 32.5 = 6 \cdot 7 - 9.5$ (check)
$U\ = 9.5$

Critical value of $U$ from Table V is 8.
Conclusion: Based on the sample data, there is no reason to suspect unequal dispersions.

EXAMPLE 18-3. Suppose that in Example 18-2 the two values of "1.3" had instead been "4.3". Then the results would have been:

| Ordered Data | Ranks |
|---|---|
| 0.2 | 1.5 |
| 0.2 | 1.5 |
| 1.9 | 5 |
| <u>2.2</u> | <u>6</u> |
| <u>3.7</u> | <u>9</u> |
| 4.3 | 10.5 |
| 4.3 | 10.5 |
| <u>4.6</u> | <u>13</u> |
| 4.8 | 12 |
| 5.1 | 7.5 |
| <u>5.1</u> | <u>7.5</u> |
| 6.3 | 4 |
| 7.6 | 3 |

## A TEST FOR DISPERSION

$n_1 = 6$
$R_1 = 56.5$
$n_2 = 7$
$R_2 = 34.5$
$U_1 = 63 - 56.5 = 6.5$
$U_2 = 70 - 34.5 = 35.5 = 42 - 6.5$ (check)
$U = 6.5$

Critical value of $U$ from Table V is 8.
Conclusion: Based on the sample data, either an extreme outcome has been observed or else the dispersions of the samples are, in fact, not equal. Notice that in this case, with only one exception, all sample values from Sample A are such that the values from Sample B are either larger or smaller.

**Exercise 18-1:** Suppose that two applicants for the position of research chemist have been interviewed and both appear equally qualified. In order to choose between them, the head of the research department gives each one ten samples of the same compound and asks them both to determine analytically the percentage of copper in their samples. The data obtained by each are as follows. Which is the better chemist, in the sense that his analyses are more precise and show less variation? (Check first whether the means can be assumed to be equal.)

Percentage of Copper

| | | | | | | | | | | |
|---|---|---|---|---|---|---|---|---|---|---|
| Applicant I | 6.3 | 6.2 | 5.8 | 5.8 | 6.0 | 5.9 | 6.1 | 6.0 | 6.1 | 6.0 |
| Applicant II | 6.5 | 5.4 | 7.2 | 5.0 | 5.9 | 6.8 | 5.9 | 6.9 | 5.0 | 5.5 |

# A TEST FOR DISPERSION

***Exercise 18–2:*** A purchaser for a major automobile company must decide between two suppliers of shock absorbers for the new models. In order to be able to predict replacement inventories of the part for service departments, he is interested in obtaining shock absorbers which yield consistent performance. He buys a random sample of eight shock absorbers from the production of each company and has them tested in the laboratory for duration of performance, recording how many hours the shock absorbers hold up before being damaged by constant use. Is the performance of one product more consistent than that of the other? (Check first whether the means can be assumed to be equal.)

*Number of Hours of Constant Use Prior to Breakdown*

| Supplier A | 563 | 558 | 570 | 562 | 565 | 553 | 575 | 560 |
|---|---|---|---|---|---|---|---|---|
| Supplier B | 564 | 572 | 402 | 35 | 760 | 575 | 568 | 692 |

# A TEST FOR DISPERSION

***Exercise 18-3:*** Two students have grades on class examinations as shown below. Are performances of the two students equally consistent?

*Grades of Two Students on Examinations*

| A | 93 | 82 | 96 | 90 | 85 | 89 | 95 |
|---|----|----|----|----|----|----|----|
| B | 93 | 70 | 98 | 77 | 75 | 97 | 99 |

**Exercise 18–4:** Several rats of two different strains were tested to learn their reaction times to a stimulus after a drug had been administered. Were the dispersions the same for the two strains of rats?

*Reaction Times of Rats After Drug*

|          |    |    |    |    |    |    |    |
|----------|----|----|----|----|----|----|----|
| Strain A | 66 | 75 | 60 | 71 | 77 | 63 | 78 |
| Strain B | 59 | 46 | 42 | 76 | 82 | 93 | 85 |

## A TEST FOR DISPERSION

**Exercise 18-5:** Using the data of Exercise 11-1 (page 95), suppose we wanted to find out whether the dispersion of the sample of tree diameters observed by the young forester was the same as that of the sample observed by the trained forester. What assumption would we violate if we used the test for dispersion discussed in this chapter?

_____

_____

**Exercise 18-6:** Using the data of Exercise 11-2 (page 96), suppose we wanted to find out whether the dispersions of the plot yields from normally cultivated and chemically controlled samples differ. What assumption would we violate if we used the test for dispersion discussed in this chapter? _____

_____

chapter 19

# MEASURES OF SKEWNESS AND KURTOSIS

Quite often, measures of central tendency such as we studied in Chapter 3 and measures of dispersion such as we studied in Chapter 17 are not sufficient to describe the properties of a frequency distribution. For example, there could be two distributions with equal means and equal standard deviations which differ considerably when graphed because one has a long "tail" extending to the left and the other has a long "tail" extending to the right. In other words, neither graph would be **symmetric.** In the graph of a symmetric distribution, that

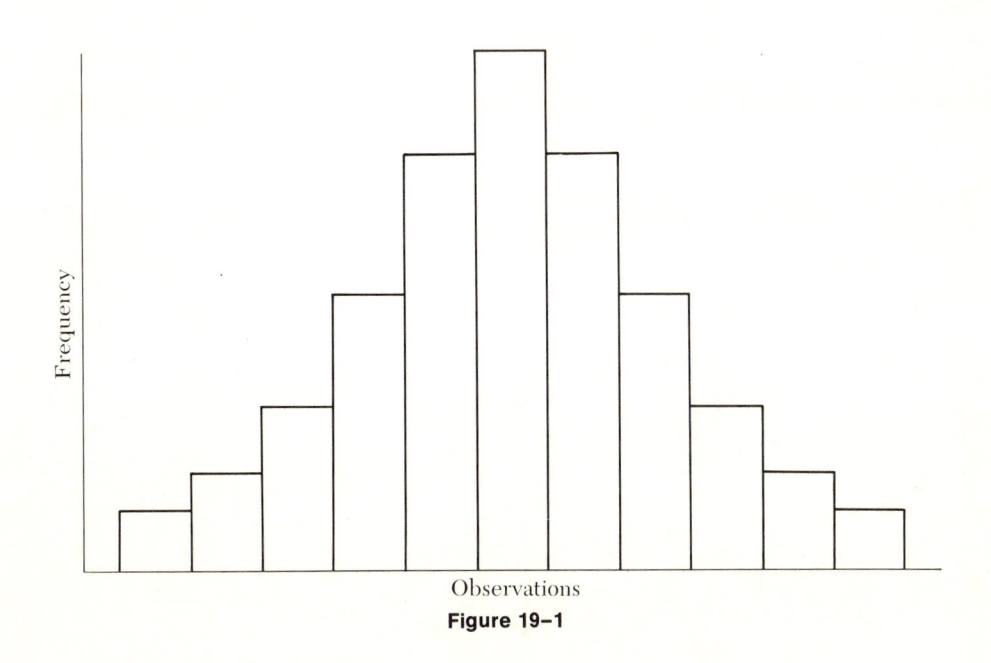

Figure 19-1

## MEASURES OF SKEWNESS AND KURTOSIS

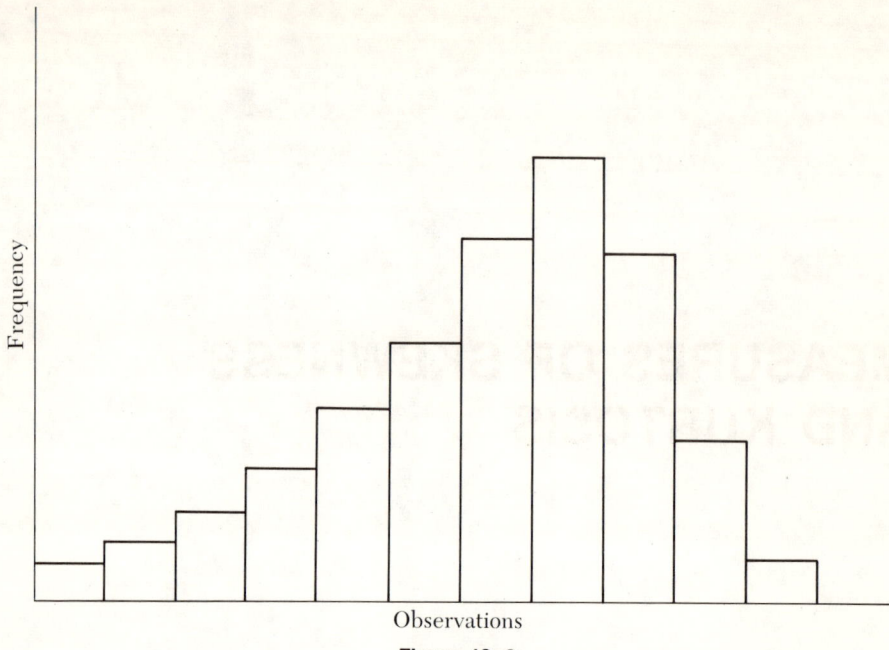

**Figure 19-2**

portion of the distribution which lies to the left of the mean is the mirror-image of the portion lying to the right of the mean, as is illustrated by Figure 19–1. A distribution that is not symmetric is said to be **skewed** or to have the property of **skewness**. If a distribution is graphed so that the measurement scale on the horizontal axis is labeled with small to large numbers from left to right, then the graph of a frequency distribution which appears to have a long "tail" on the left, such as that in Figure 19–2, is said to be **left-skewed** or negatively

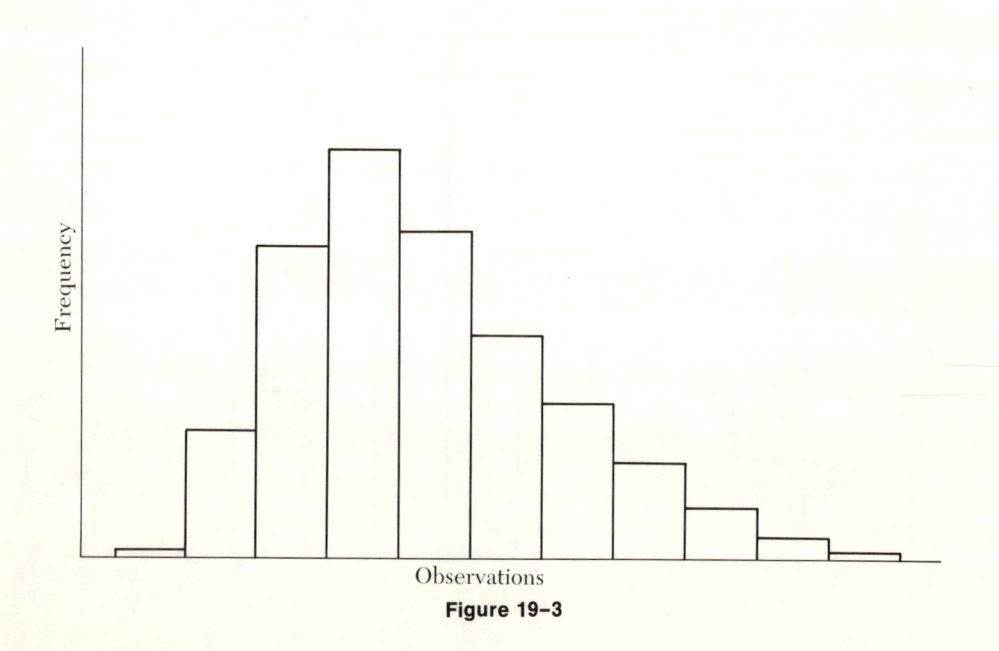

**Figure 19-3**

## MEASURES OF SKEWNESS AND KURTOSIS

skewed. If the "tail" appears on the right, as in Figure 19–3, it is said to be **right-skewed** or positively skewed. After calculating the measures of location and dispersion of a distribution, classifying it as being symmetric, left-skewed, or right-skewed serves to describe further its basic properties.

However, there is one further property which is needed to describe adequately the basic properties of the distribution: we must indicate whether the graph of the distribution appears flat or peaked. In describing the degree of **kurtosis** or peakedness of a distribution, a comparison is usually made between the distribution in question and the *normal curve*, which is discussed more fully in Chapter 26. Any curve which has the same degree of kurtosis as the standard normal

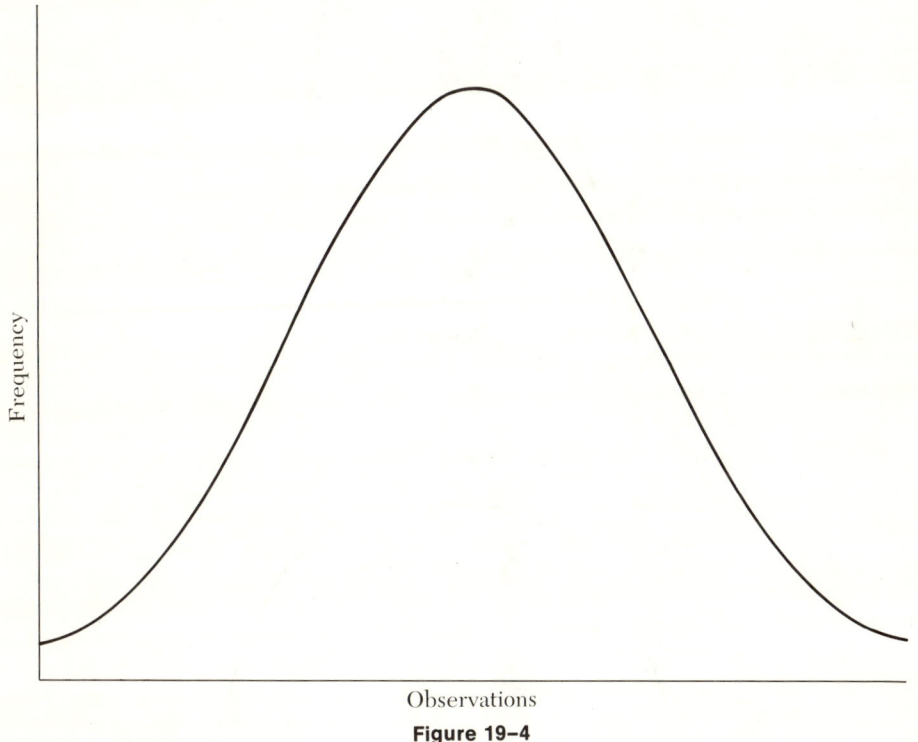

**Figure 19–4**

curve is said to be **mesokurtic.** For an illustration of a mesokurtic curve, see Figure 19–4. One which is more peaked than the standard normal curve—it seems to leap up!—is said to be **leptokurtic**, as in Figure 19–5, while one which is less peaked than the standard normal curve is said to be **platykurtic.** A platykurtic curve seems to have a plateau, as is shown by Figure 19–6.

The degree of skewness and kurtosis is usually easily observed in the graph of a frequency distribution, regardless of whether a histo-

## MEASURES OF SKEWNESS AND KURTOSIS

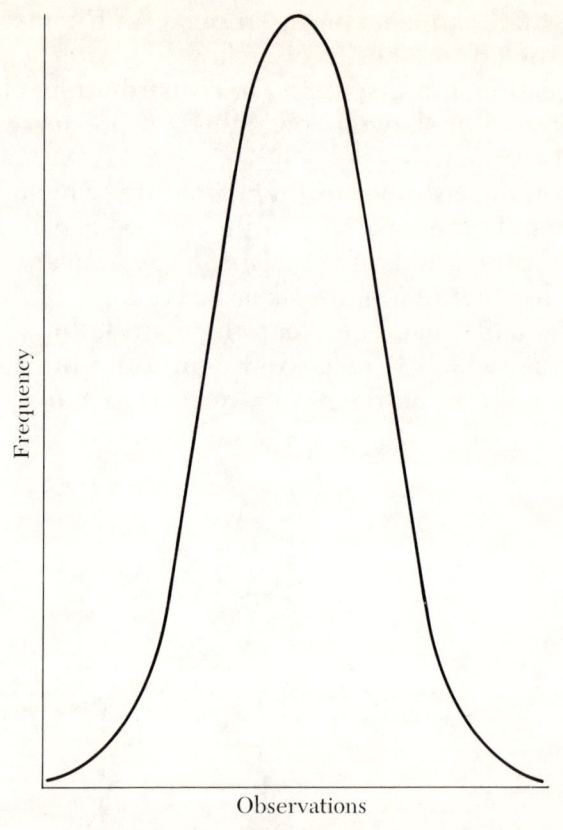

**Figure 19–5**

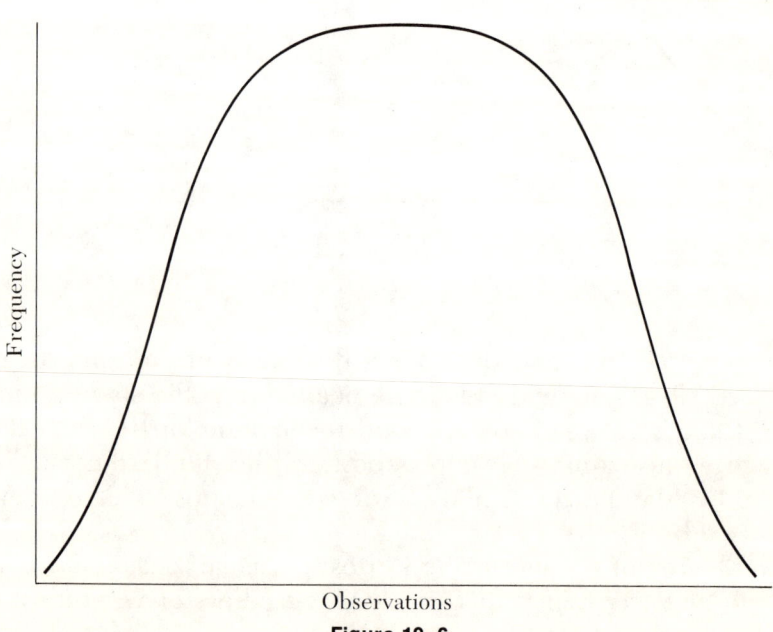

**Figure 19–6**

## MEASURES OF SKEWNESS AND KURTOSIS

gram is constructed—as in Figures 19-1, 19-2, and 19-3—or a smooth curve, illustrated by Figures 19-4, 19-5, and 19-6.

In this chapter, along with Chapters 3 and 17, we have considered the four basic properties which may be used to describe fully any frequency distribution: central tendency, dispersion, skewness, and kurtosis. Whenever descriptive techniques are desired, these four properties should be included in the description and some measure calculated for each. The **sample moments** of a distribution are quite useful in providing mathematical measures of these four properties. The first four sample moments and the first four central sample moments are given in Table 19-1.

**TABLE 19-1** A LISTING OF THE FIRST FOUR SAMPLE MOMENTS AND THE FIRST FOUR CENTRAL SAMPLE MOMENTS.

| k | Sample Moments | Central Sample Moments |
|---|---|---|
| 1 | $m_1' = \dfrac{\Sigma X}{n}$ | $m_1 = \dfrac{\Sigma (X - \bar{x})}{n}$ |
| 2 | $m_2' = \dfrac{\Sigma X^2}{n}$ | $m_2 = \dfrac{\Sigma (X - \bar{x})^2}{n}$ |
| 3 | $m_3' = \dfrac{\Sigma X^3}{n}$ | $m_3 = \dfrac{\Sigma (X - \bar{x})^3}{n}$ |
| 4 | $m_4' = \dfrac{\Sigma X^4}{n}$ | $m_4 = \dfrac{\Sigma (X - \bar{x})^4}{n}$ |

The $k$th sample moment is given by

$$m_k' = \frac{\Sigma X^k}{n},$$

and the $k$th central sample moment is given by

$$m_k = \frac{\Sigma (X - \bar{x})^k}{n}.$$

We shall see as we define measures of skewness and kurtosis, and as we review the most used measures of central tendency and dispersion, that the four basic properties necessary to describe fully any given frequency distribution may be obtained using the first four moments or central moments.

EXAMPLE 19-1. From Table 19-1, we may calculate any of the moments or central moments for any set of data. Suppose that we had observations 1, 1, 3, 2, 4, 1. The sample moments and central sample moments for this set of data are calculated in Table 19-2.

# MEASURES OF SKEWNESS AND KURTOSIS

**TABLE 19-2** MOMENTS AND CENTRAL MOMENTS FOR DATA OF EXAMPLE 19-1.

| k | Sample Moments | Central Sample Moments |
|---|---|---|
| 1 | $m_1' = \dfrac{1+1+3+2+4+1}{6} = 2$ | $m_1 = \dfrac{-1-1+1+0+2-1}{6} = 0$ |
| 2 | $m_2' = \dfrac{1+1+9+4+16+1}{6} = \dfrac{32}{6} = 5.3$ | $m_2 = \dfrac{1+1+1+0+4+1}{6} = \dfrac{8}{6} = 1.3$ |
| 3 | $m_3' = \dfrac{1+1+27+8+64+1}{6} = \dfrac{102}{6} = 17$ | $m_3 = \dfrac{-1-1+1+0+8-1}{6} = \dfrac{6}{6} = 1$ |
| 4 | $m_4' = \dfrac{1+1+81+16+256+1}{6} = \dfrac{356}{6} = 59.3$ | $m_4 = \dfrac{1+1+1+0+16+1}{6} = \dfrac{20}{6} = 3.3$ |

Our most often used measure of central tendency, the mean, is given by the first sample moment:

$$\bar{x} = \frac{\Sigma X}{n} = m_1'.$$

Our most often used measure of dispersion, the standard deviation, is the square root of the variance; the variance, in turn, is a multiple of the second central sample moment:

$$s^2 = \frac{\Sigma(X-\bar{x})^2}{n-1} = \left(\frac{n}{n-1}\right)\left(\frac{\Sigma(X-\bar{x})^2}{n}\right) = \frac{n}{n-1}\, m_2.$$

One of the common measures of skewness involves the third central moment and the third power of the second central moment; it is often designated $\alpha_3$:

$$\alpha_3 = \frac{m_3}{\sqrt{m_2^3}} = \frac{\dfrac{\Sigma(X-\bar{x})^3}{n}}{\sqrt{\left[\dfrac{\Sigma(X-\bar{x})^2}{n}\right]^3}} = \frac{\sqrt{n}\,\Sigma(X-\bar{x})^3}{\sqrt{[\Sigma(X-\bar{x})^2]^3}}.$$

Another commonly used measure of skewness, called the Pearsonian coefficient of skewness, can be found by the formula:

$$Sk = \frac{3(\text{mean}-\text{median})}{\text{standard deviation}}.$$

EXAMPLE 19-2. Using the same data as Example 19-1 (1,1,3,2,4,1), we can calculate

$$\alpha_3 = \frac{2}{\sqrt{1.3^3}} = 1.35$$

## MEASURES OF SKEWNESS AND KURTOSIS

and

$$Sk = \frac{3(2 - 1.5)}{\sqrt{1.6}} = 1.19.$$

Both measures indicate that the distribution is positively skewed.

The Pearsonian coefficient of skewness brings out the relation between the mean and the median in a skewed distribution which has exactly one mode, or is *unimodal*. Recall from Chapter 3 that the mean is the center of gravity, the mode is the high point, and the median divides the area in half. Obviously, then, in a symmetric unimodal distribution the mean, median, and mode are equal. In a left-skewed distribution the mean is less than the median, which in turn is less than the mode; and in a right-skewed distribution, the mean is greater than the median, which is greater than the mode. These relationships are illustrated by Figures 19–7 and 19–8. Thus, the quantity "mean minus median" results in a negative number for a left-skewed curve and in a positive number for a right-skewed curve. Similarly, there will be larger negative values of $X - \bar{x}$ than positive values if we have a left-skewed curve, so that when they are cubed and summed for $\alpha_3$, a negative value results. By a similar argument, a positive value of $\alpha_3$ results for a right-skewed distribution. With both of these measures of skewness we would expect a symmetric curve to have $\alpha_3 = Sk = 0$, a negatively skewed curve to have a negative value

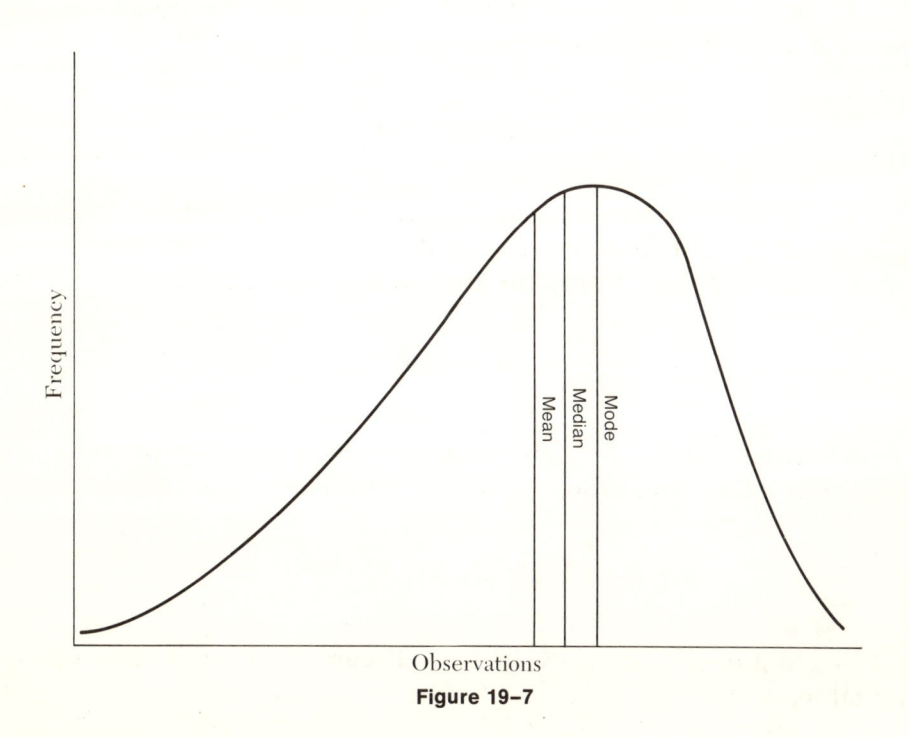

**Figure 19–7**

## MEASURES OF SKEWNESS AND KURTOSIS

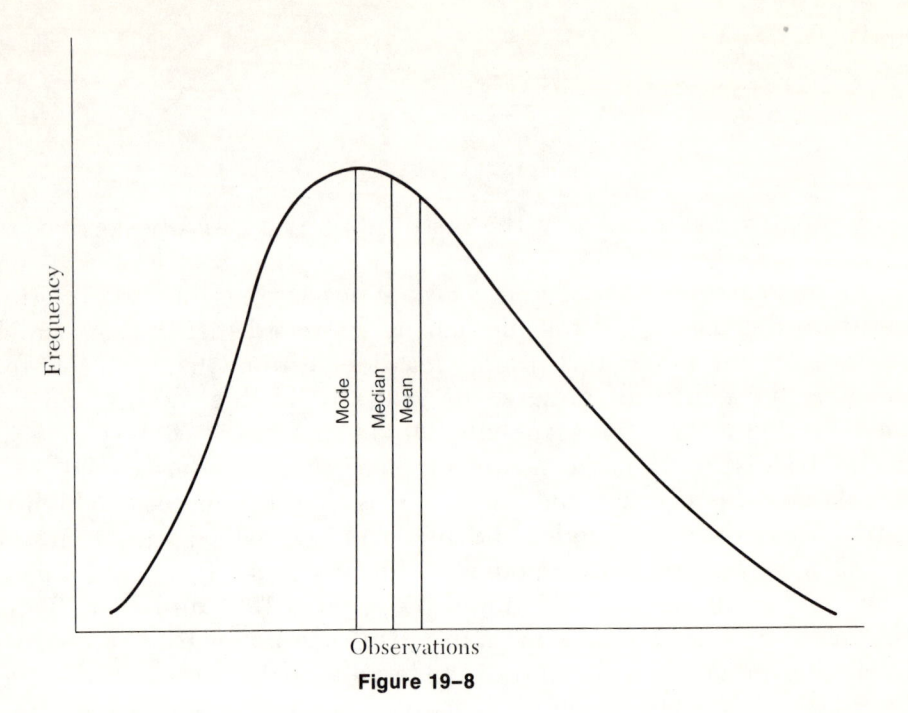

**Figure 19–8**

for both $\alpha_3$ and $Sk$, and a positively skewed curve to have a positive value for $\alpha_3$ and $Sk$. Minor deviations from this rule sometimes occur when the median is calculated from data with repeated values or from grouped data.

A measure of kurtosis, called $\alpha_4$, involves the fourth central moment and the square of the second central moment:

$$\alpha_4 = \frac{m_4}{m_2^2} - 3.$$

For a mesokurtic distribution $\alpha_4$ is zero; for a platykurtic distribution $\alpha_4$ is negative; and for a leptokurtic distribution $\alpha_4$ is positive.

EXAMPLE 19–3. Using the data of Example 19–1 again,

$$\alpha_4 = \frac{3.3}{(1.3)^2} - 3 = -1.05 \text{ (platykurtic)}.$$

EXAMPLE 19–4. Suppose that we wish to describe fully the properties of the distribution of the following set of observations:

$$24, 30, 27, 36, 28, 34, 32, 29, 31, 29$$

Let us first group the data, as shown in Figure 19–9 and its accompanying table.

## MEASURES OF SKEWNESS AND KURTOSIS

**Frequency Table**

| RANGE | FREQUENCY |
|---|---|
| 24–26 | 1 |
| 27–29 | 4 |
| 30–32 | 3 |
| 33–35 | 1 |
| 36–38 | 1 |

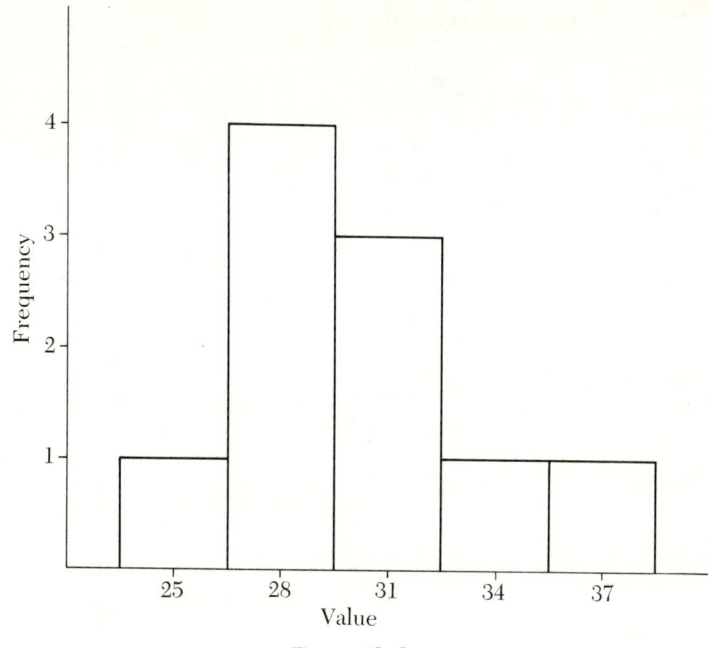

Figure 19-9

The mean is

$$\bar{x} = \frac{\Sigma X}{n} = m_1' = 30$$

and the median is 29.5. Table 19-3 shows the deviations from the mean, their powers, and the sums needed to calculate the central sample moments.

**TABLE 19-3  CALCULATIONS FOR EXAMPLE 19-4.**

| $X$ | $X - \bar{x}$ | $(X - \bar{x})^2$ | $(X - \bar{x})^3$ | $(X - \bar{x})^4$ |
|---|---|---|---|---|
| 24 | −6 | 36 | −216 | 1296 |
| 27 | −3 | 9 | −27 | 81 |
| 28 | −2 | 4 | −8 | 16 |
| 29 | −1 | 1 | −1 | 1 |
| 29 | −1 | 1 | −1 | 1 |
| 30 | 0 | 0 | 0 | 0 |
| 31 | 1 | 1 | 1 | 1 |
| 32 | 2 | 4 | 8 | 16 |
| 34 | 4 | 16 | 64 | 256 |
| 36 | 6 | 36 | 216 | 1296 |
|  | $\Sigma(X - \bar{x}) = 0$ | $\Sigma(X - \bar{x})^2 = 108$ | $\Sigma(X - \bar{x})^3 = 36$ | $\Sigma(X - \bar{x})^4 = 2964$ |

## MEASURES OF SKEWNESS AND KURTOSIS

The central sample moments are then

$$m_1 = 0; \quad m_2 = \frac{108}{10} = 10.8; \quad m_3 = \frac{36}{10} = 3.6; \quad m_4 = \frac{2964}{10} = 296.4.$$

From these we may calculate the standard deviation and $\alpha_3$:

$$\text{Std. dev.} = \sqrt{\frac{\Sigma(X-\bar{x})^2}{n-1}} = \sqrt{\frac{n}{n-1} m_2} = \sqrt{\frac{108}{9}} = \sqrt{12} = 3.46$$

$$\alpha_3 = \frac{3.6}{10.8\sqrt{10.8}} = \frac{3.6}{10.8(3.29)} = 0.10 \text{ (positively skewed)}$$

$$\text{Pearsonian Coefficient of Skewness} = \frac{3(30-29.5)}{3.46} = \frac{1.5}{3.46} = 0.43$$

$$\alpha_4 = \frac{296.4}{(10.8)^2} - 3 = -0.46 \text{ (platykurtic)}$$

Thus we see that the distribution tends to center near the value 30, its measure of dispersion is 3.46, and it is positively skewed and platykurtic.

**Exercise 19-1:** Find the mean, the median, the standard deviation, $m_2$, $m_3$, $m_4$, $Sk$, $\alpha_3$, and $\alpha_4$ for the following data and then give a brief description of the data in words: 2, 3, 0, 1, 5, 7, 2, 3, 4, 3.

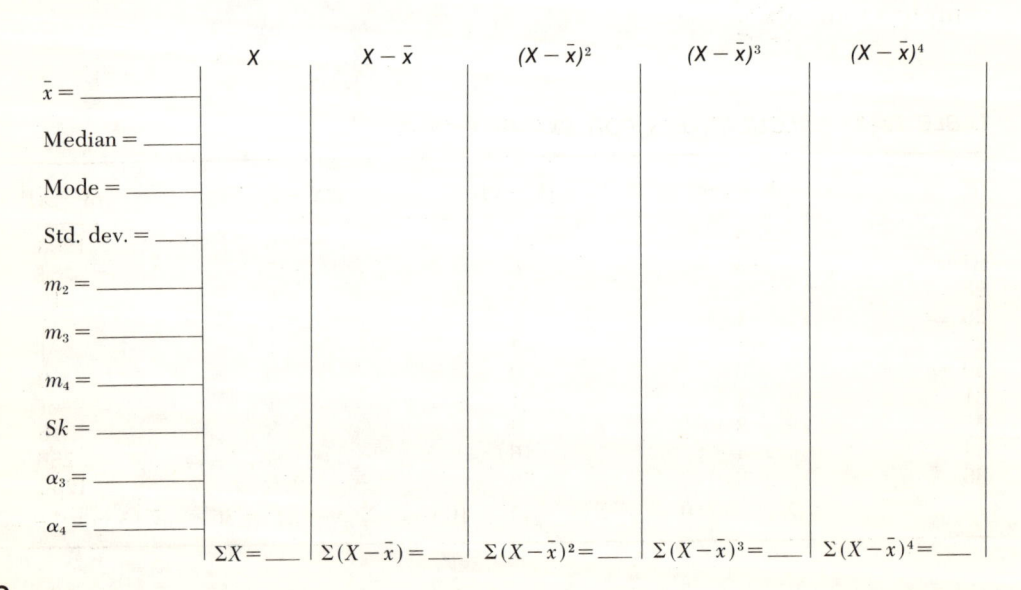

| | X | X − x̄ | (X − x̄)² | (X − x̄)³ | (X − x̄)⁴ |
|---|---|---|---|---|---|
| $\bar{x}$ = \_\_\_ | | | | | |
| Median = \_\_\_ | | | | | |
| Mode = \_\_\_ | | | | | |
| Std. dev. = \_\_\_ | | | | | |
| $m_2$ = \_\_\_ | | | | | |
| $m_3$ = \_\_\_ | | | | | |
| $m_4$ = \_\_\_ | | | | | |
| $Sk$ = \_\_\_ | | | | | |
| $\alpha_3$ = \_\_\_ | | | | | |
| $\alpha_4$ = \_\_\_ | | | | | |
| | $\Sigma X$=\_\_ | $\Sigma(X-\bar{x})$=\_\_ | $\Sigma(X-\bar{x})^2$=\_\_ | $\Sigma(X-\bar{x})^3$=\_\_ | $\Sigma(X-\bar{x})^4$=\_\_ |

## MEASURES OF SKEWNESS AND KURTOSIS

Description: _____

_____

_____

**Exercise 19-2:** Find the mean, median, standard deviation, $m_2$, $m_3$, $m_4$, $Sk$, $\alpha_3$, and $\alpha_4$ for the following data and then give a brief description of the data in words: 0, 1, 5, 2, 6, 5, 4, 7, 6, 4.

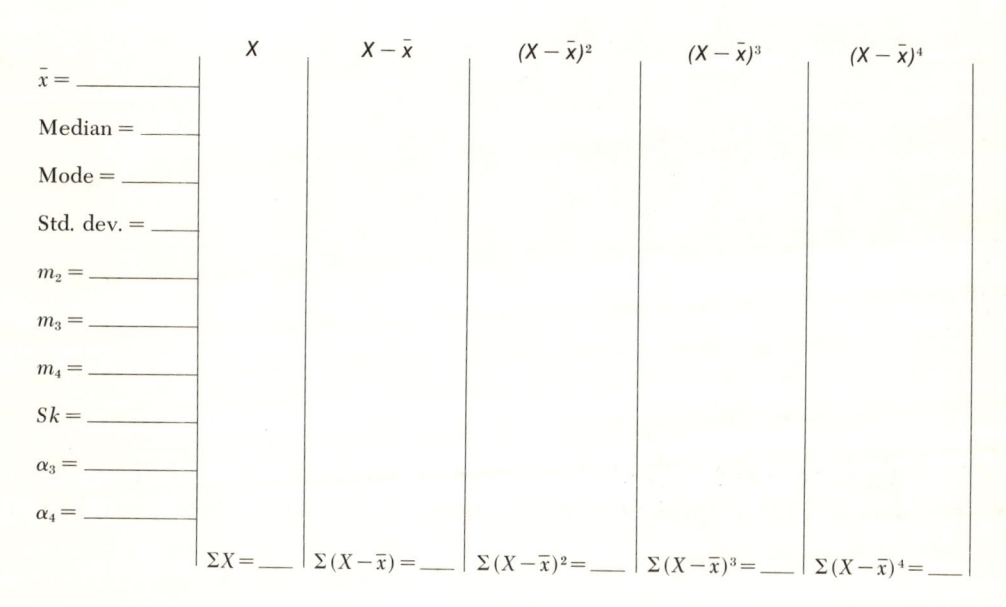

| | X | $X - \bar{x}$ | $(X - \bar{x})^2$ | $(X - \bar{x})^3$ | $(X - \bar{x})^4$ |
|---|---|---|---|---|---|
| $\bar{x} =$ ___ | | | | | |
| Median = ___ | | | | | |
| Mode = ___ | | | | | |
| Std. dev. = ___ | | | | | |
| $m_2 =$ ___ | | | | | |
| $m_3 =$ ___ | | | | | |
| $m_4 =$ ___ | | | | | |
| $Sk =$ ___ | | | | | |
| $\alpha_3 =$ ___ | | | | | |
| $\alpha_4 =$ ___ | | | | | |
| | $\Sigma X =$ ___ | $\Sigma(X-\bar{x}) =$ ___ | $\Sigma(X-\bar{x})^2 =$ ___ | $\Sigma(X-\bar{x})^3 =$ ___ | $\Sigma(X-\bar{x})^4 =$ ___ |

Description: _____

_____

_____

**Exercise 19-3:** Find the necessary measures to describe fully the data of both groups of students for Exercise 18-3, page 154.

**MEASURES OF SKEWNESS AND KURTOSIS**

*Exercise 19-4:* Find the necessary measures to describe fully the data of both groups of reaction times for Exercise 18-4, page 155.

# MEASURES OF SKEWNESS AND KURTOSIS

*TERMS TO REMEMBER:*

**Symmetric:** The quality, in a frequency distribution graph, of having the same appearance both to the left and to the right of the mean.

**Skewed:** A distribution that is not symmetric is skewed.

**Skewness:** The quality, in a frequency distribution graph, of not having the same appearance both to the left and to the right of the mean.

**Left-skewed, or Negatively Skewed:** Several outlying observations have smaller numerical value and very few outliers have larger numerical value than the majority of the observations. The graph has a tail stretching out toward the left.

**Right-skewed, or Positively Skewed:** More large-valued outliers than small-valued outliers are observed. The graph has a tail stretching out toward the right.

**Kurtosis:** The term which describes the appearance of peakedness or flatness in a frequency distribution curve.

**Mesokurtic:** Having the same kurtosis as the standard normal curve.

**Leptokurtic:** Having a more peaked appearance than the normal curve.

**Platykurtic:** Having a flatter appearance than the normal curve.

**Sample Moments:** Quantities, calculated from sample data, that can be used to measure the central tendency, dispersion, skewness, and kurtosis of a frequency distribution. The $k$th sample moment is given by $m_k' = \frac{\Sigma X^k}{n}$. The $k$th central sample moment is given by $m_k = \frac{\Sigma (X - \bar{x})^k}{n}$, where $\bar{x}$ is the mean of the $n$ sample observations $X$.

chapter 20

# THE WALD-WOLFOWITZ RUNS TEST

In Chapter 6 we considered a one-sample runs test. Its primary purpose was to determine whether a sequence of observations was in fact random in time or space. In Chapters 11, 12, and 16 we considered tests designed to indicate whether two samples differed significantly in their measures of central tendency. If they did, then we know that the samples were drawn from different populations. In Chapter 18 we applied a test to determine whether two samples came from populations which differed in their measures of dispersion. Finally, in Chapter 19 we found that two samples may also differ significantly in their measures of skewness and kurtosis, again implying that they came from different populations. Now we shall consider a test which combines all of these concepts.

Using the number of runs observed in the combined observations from both samples, the Wald-Wolfowitz Runs Test determines whether the two samples' means, dispersions, degrees of skewness, or kurtosis differ to such an extent that the differences are too large to be caused by random fluctuations or sampling error. If the samples differ significantly in any combination of these four characteristics, then the samples were drawn from different populations. In this chapter we shall define a run as a sequence of observations from one sample that is preceded and succeeded by observations from another sample or else by no observations at all.

EXAMPLE 20–1. Suppose we had the following two sets of observations as our experimental data:

# THE WALD-WOLFOWITZ RUNS TEST

A:  3,  5,  6,  6,  7,  10

B:  1,  2,  8,  8,  9,  13.

Suppose we ordered them and underlined those observations from A:

1, 2, 3, 5, 6, 6, 7, 8, 8, 9, 10, 13

Then we would have 5 runs. Call $n_1$ the number of observations from Sample A, and $n_2$ the number from Sample B. Then $n_1 = n_2 = 6$. From Table VII of the appendix, we see that 5 runs are not unusual if in fact the two samples were drawn from the same population. Three or fewer runs would have implied that either we had observed an event expected to occur less than 5% of the time, or else the samples came from different populations.

The following simple steps are all that are necessary to apply the Wald-Wolfowitz Runs Test.
 (1) Combine the two sets of observations in a single ordered series, either ascending or descending. Identify the observations as to the group to which they belong. A sequence of observations from one sample that is preceded and succeeded by an observation from the other sample or by nothing at all constitutes a run.
 (2) Count the number of runs and specify $n_1$ and $n_2$, the two sample sizes.
 (3) Compare the observed number of runs with the critical value given in Table VII for the given sample sizes. The observed number of runs in the sequence of combined observations will be equal to or smaller than the critical value less than 5% of the time if in fact there is no true difference of any kind (in central tendency, dispersion, skewness, or kurtosis) between the samples.

The next two examples we shall consider in order to illustrate this technique will be the same examples considered in Chapter 18. Also, the exercises will involve the same sample data that was considered in Chapter 18. The purpose of using the same sample data for different tests both here and elsewhere in this text is for the student to work with the same data enough to see the relationships among the observations and notice those patterns which illustrate the basic properties of their distributions. By using the same sample data, we do NOT mean to imply that the experimenter should apply several tests to his data and see which one gives the "nicest" results. In actual experimentation, one must decide upon the appropriate statistical procedure **BEFORE THE EXPERIMENT IS PERFORMED** and

## THE WALD-WOLFOWITZ RUNS TEST

then apply only that one single test. Trying to "shop around" and find a statistical procedure that will prove your point is—plainly speaking—unethical and invalid!

EXAMPLE 20-2. (Using same data as Example 18-1, page 149):
1. *Ordered data:* <u>29 30</u> <u>32</u> <u>35 36 37 38 39 42 43 45 49 56</u>
2. *# Runs* = 11    $n_1 = 7$    $n_2 = 6$
3. *Critical value* = 4. Since 11 is larger than 4, the conclusion is that these two samples came from the same population. In other words, the observed results indicate that the actual differences are caused by sampling error and do not imply a true difference between the populations sampled.

EXAMPLE 20-3. (Using same data as Example 18-2, page 150):
1. *Ordered data:* <u>0.2   0.2</u>   <u>1.3   1.3</u>   <u>1.9</u>   <u>2.2   3.7   4.6   4.8   5.1   5.1</u>   <u>6.3   7.6</u>.
2. *# Runs* = 7   $n_1 = 6$   $n_2 = 7$
3. *Critical value* = 4.   Since 7 is larger than 4, the conclusion is that the sample evidence would lead us to believe that these two samples came from the same population.

Consider next why only the presence of too few runs implies that the samples differ. If two samples come from populations that differ in their means and no other way, then one would expect most of the observations from one sample to be larger than those from the other samples, yielding possibly as few as two runs. Suppose two samples came from populations that differ only in their dispersions. Then one would expect the observations from the less dispersed sample to be "bunched up" or clumped in the center of those from the more dispersed sample, yielding possibly as few as three runs. If they differed in skewness, there would be a clumping of one sample on one end and a clumping of the other on the other end. Finally, if they differed in kurtosis, we would expect a pattern of clumping similar to that for differences of dispersion. All these conditions, when present, result in a small number of runs. On the other hand, two samples drawn from the same population would have their observations fairly evenly mixed together, so that the sequence of combined observations would exhibit a comparatively large number of runs, even though we would not expect complete alternation, like I, II, I, II, and so on.

**Exercise 20-1:** Using the Wald-Wolfowitz test, find whether the analyses of the two applicants in Exercise 18-1, page 152, came from the same population.

**Exercise 20-2:** Using the Wald-Wolfowitz test, find whether the number of hours of constant use prior to breakdown is from the same population for the two suppliers in Exercise 18-2, page 153.

**Exercise 20-3:** Test whether the grades of the two students are from the same population in Exercise 18-3, page 154.

**Exercise 20-4:** Test whether the two sets of reaction times in Exercise 18-4, page 155, can be assumed to come from the same population.

chapter 21

# THE POISSON DISTRIBUTION AND ITS APPLICATIONS

**THE DISTRIBUTION OF RARE OR ACCIDENTAL EVENTS**

Many occurrences in our everyday experience follow an "accidental" type of pattern—such as the pattern of flying bomb hits in London in World War II, or the instances of German cavalrymen who died between 1875 and 1894 after being kicked by their horses, or the typing mistakes made by a secretary, or breakdowns of an electronic computer, or the pattern of radioactive emanations of a radium source, or the annual number of suicides in the United States, or the number of bacteria on a petri plate, or the people arriving to be served at a table in a university registration line, or the incoming telephone calls at a switchboard, or the ships arriving at a seaport, or the number of raisins or chocolate chips in a cookie. Accidental occurrences such as these are called Poisson events because they have a probability distribution which was first studied by the French mathematician, S. D. Poisson, in the early 1800's.

Poisson events have three basic properties in common: (1) the events in one time or space interval do not affect those in any other time or space interval; (2) any given time or space interval may be subdivided successively until it contains only one event; (3) the conditions of the experiment remain constant over time or over space.

EXAMPLE 21–1. To illustrate these basic properties, consider

## THE POISSON DISTRIBUTION AND ITS APPLICATIONS

the number of incoming telephone calls on a switchboard of a major corporation in the time interval between 10 A.M. and 11 A.M. Obviously, one person making a phone call to this corporation is completely unaffected by whether or not another person is also making or will make a phone call to that corporation at the same time or at any subsequent time during this hour-long period; thus property 1 is satisfied. If we subdivide the hour into minutes, and, if necessary, the minutes into seconds or tenths of a second, or even smaller periods, we may conceivably isolate any one phone call within a given period of time; there would be no intervals of time which contained so many calls that we could not subdivide the intervals and isolate each call, if necessary, so that property 2 is also satisfied. Finally, we would expect to have as many phone calls at any one time as at any other within the specified hour, so property 3 is satisfied.

As an example of a Poisson event over a spatial interval, consider a raindrop falling on a square of sidewalk. Its position within the square is unaffected by that of any other raindrop. If we observe the raindrops and mark their positions for a very short time, we can subdivide the square into smaller and smaller squares until each raindrop is isolated. Furthermore, there is no reason to suspect that the raindrops will all fall in any one place or that they will change their pattern of falling because of positions taken by former or subsequent fallen raindrops.

EXAMPLE 21-2. Suppose we are eating chocolate chip cookies which have an average of two chocolate chips per cookie. If we count the number of chips in the first ten cookies that we eat, we might

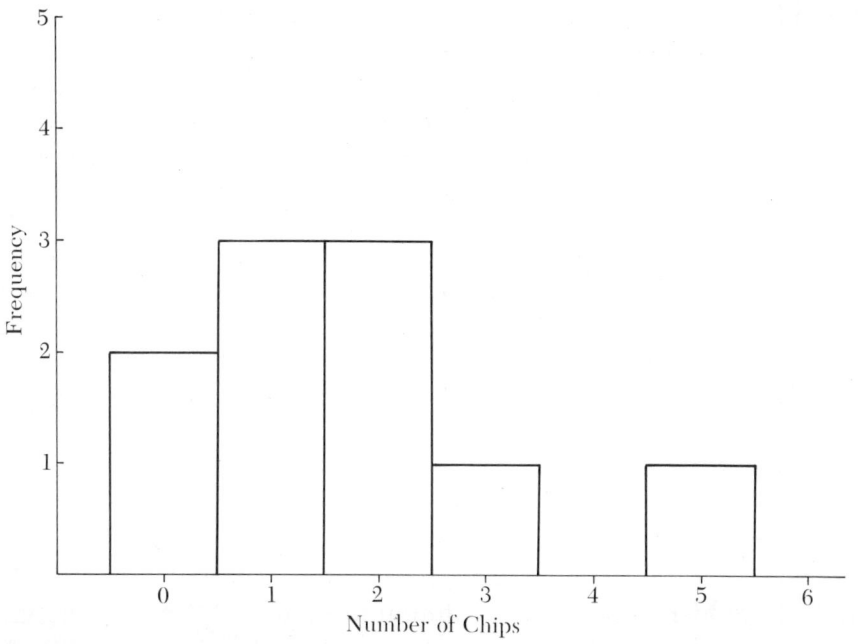

**Figure 21-1** Number of chocolate chips in ten cookies. Example of an observed Poisson distribution with a mean of 2.

## THE POISSON DISTRIBUTION AND ITS APPLICATIONS

observe 1, 0, 3, 2, 1, 1, 2, 0, 2, 5 chips respectively. We could record the observations in a simple table as follows:

|  | Observations |  |  |  |  |  |  |  |  |  |
|---|---|---|---|---|---|---|---|---|---|---|
| Experiment No.:<br>(i.e., cookie no.) | 1 | 2 | 3 | 4 | 5 | 6 | 7 | 8 | 9 | 10 |
| No. of chips observed: | 1 | 0 | 3 | 2 | 1 | 1 | 2 | 0 | 2 | 5 |

The sample mean would be 1.7, and a histogram of the data would look like Figure 21-1.

In Figure 21-2 we may compare the histograms of theoretical Poisson distributions with means of 1, 2, and 5. The second of these is the model for our sample distribution of Example 21-1. From it one may devise a sampling urn which has a Poisson distribution with mean of 2 by using numbered beads or discs with the following distribution of numbers: 27 zeros, 54 ones, 54 twos, 36 threes, 18 fours, 8 fives, 2 sixes, and 1 seven (or we may use only half as many discs numbered with 14 zeros, 27 ones, 27 twos, 18 threes, 9 fours, 4 fives, and 1 six).

***Exercise 21-1:*** Using a sampling device for the Poisson distribution with mean 2, take 10 samples, each with 10 observations in the following manner:

(1) Draw 10 numbers randomly from your sampling device and record them as the 10 observations for Experiment Number 1 in Table 21-1. Replace the numbers. (2) Draw 10 more numbers and record them as Experiment 2 in the table. (3) Repeat until 10 experiments are completed. (4) Next calculate the mean for each sample and record it in the table. (5) Draw a frequency *histogram* for each sample in the space provided. (6) Then combine the data into one sample of 100 observations, drawing a histogram of the outcomes for the combined experiment.

Step 6 can be most easily performed by reading the frequencies of "0" from each histogram, adding them, and plotting the frequency of "0" on the combined histogram; then doing the same for "1", "2", and so on. Calculate the mean for the combined experiment. Is it closer to the expected mean of "2" than those of the individual experiments? _____ Why or why not? _____

_____

Which of the histograms is most similar to Figure 21-2? _____

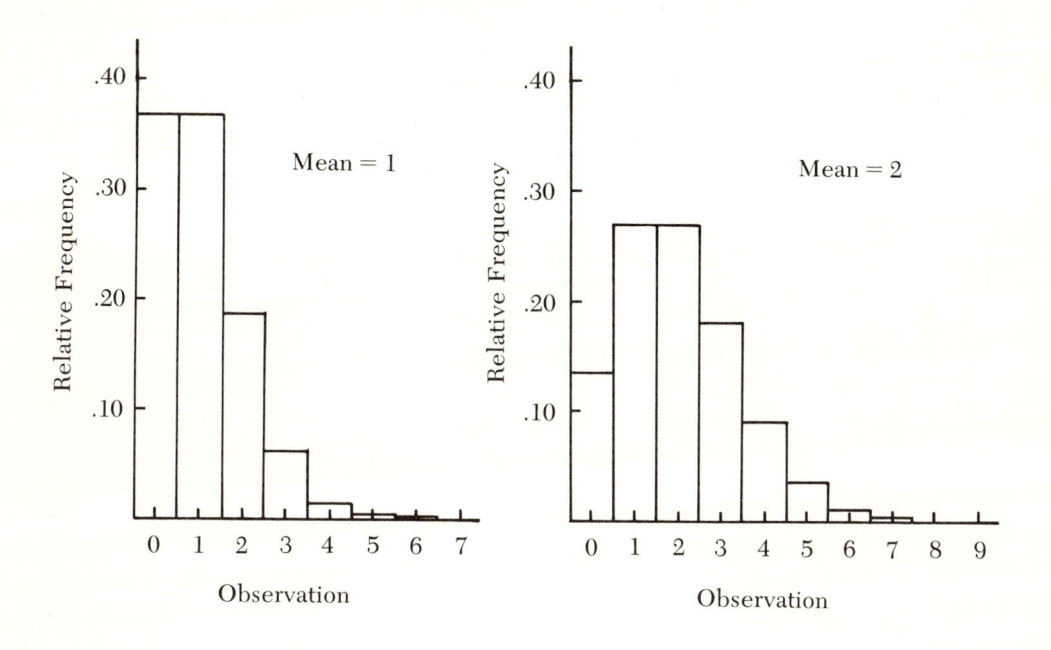

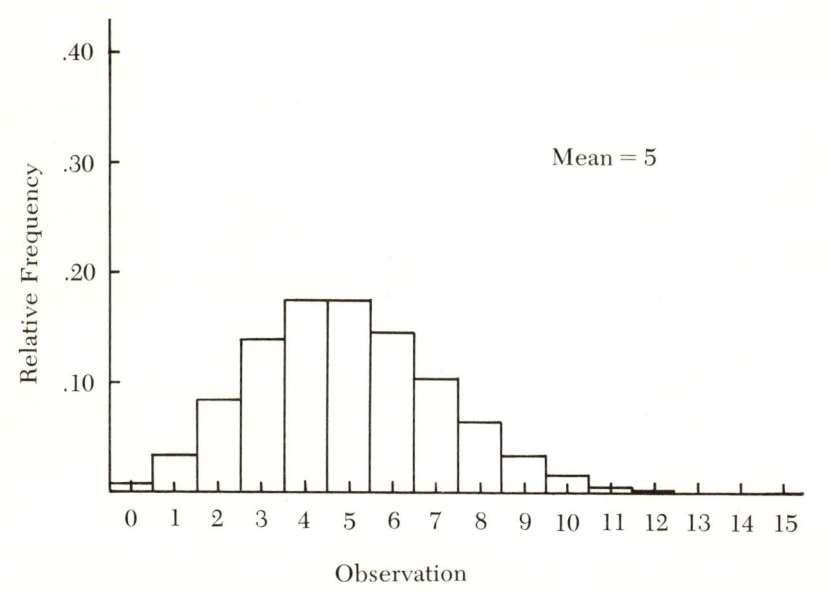

**Figure 21-2.** Histograms of Poisson distributions which have means of 1, 2, and 5.

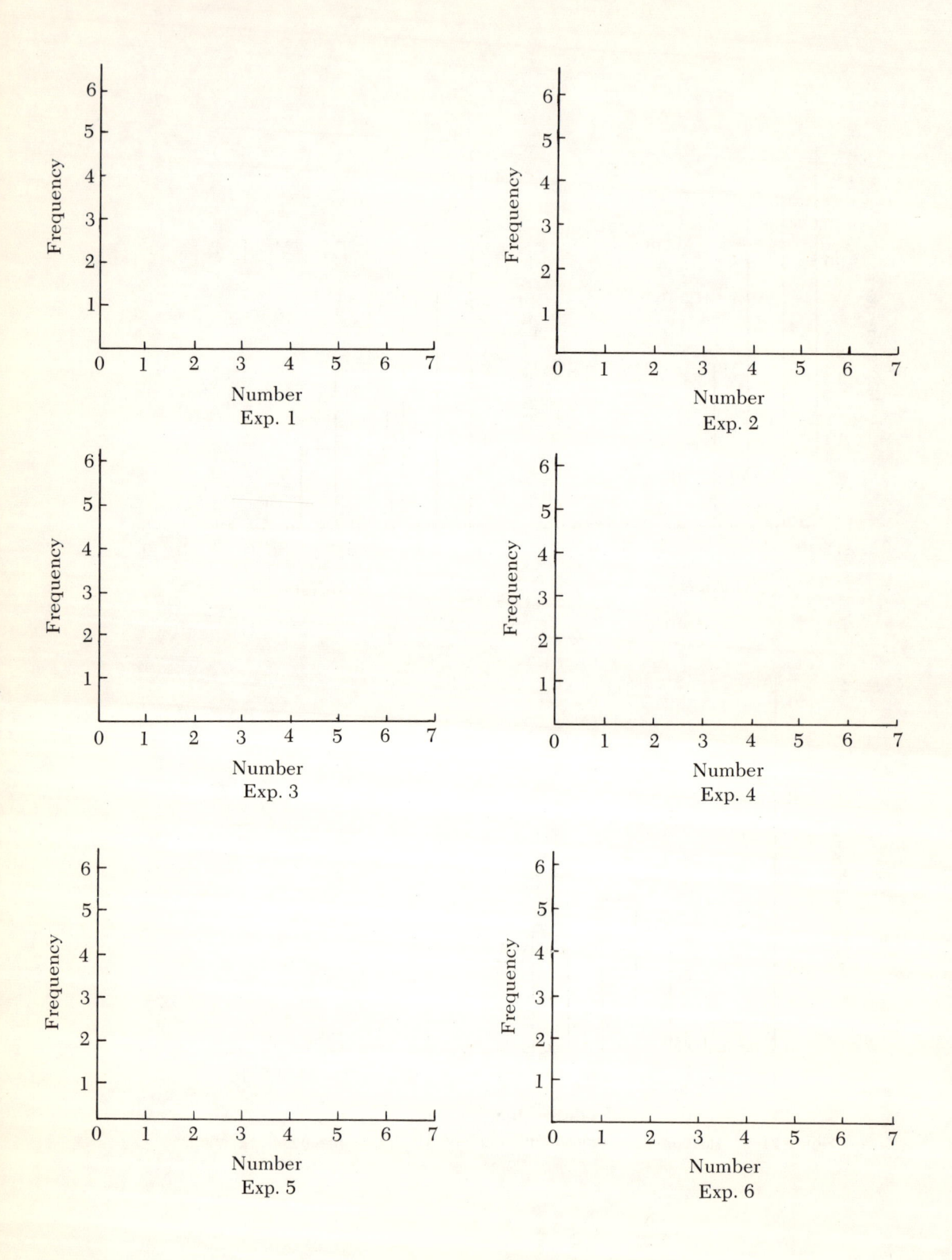

# THE POISSON DISTRIBUTION AND ITS APPLICATIONS

Exp. 1

Exp. 2

Exp. 3

Exp. 4

Exp. 5

Exp. 6

Number
Exp. 7

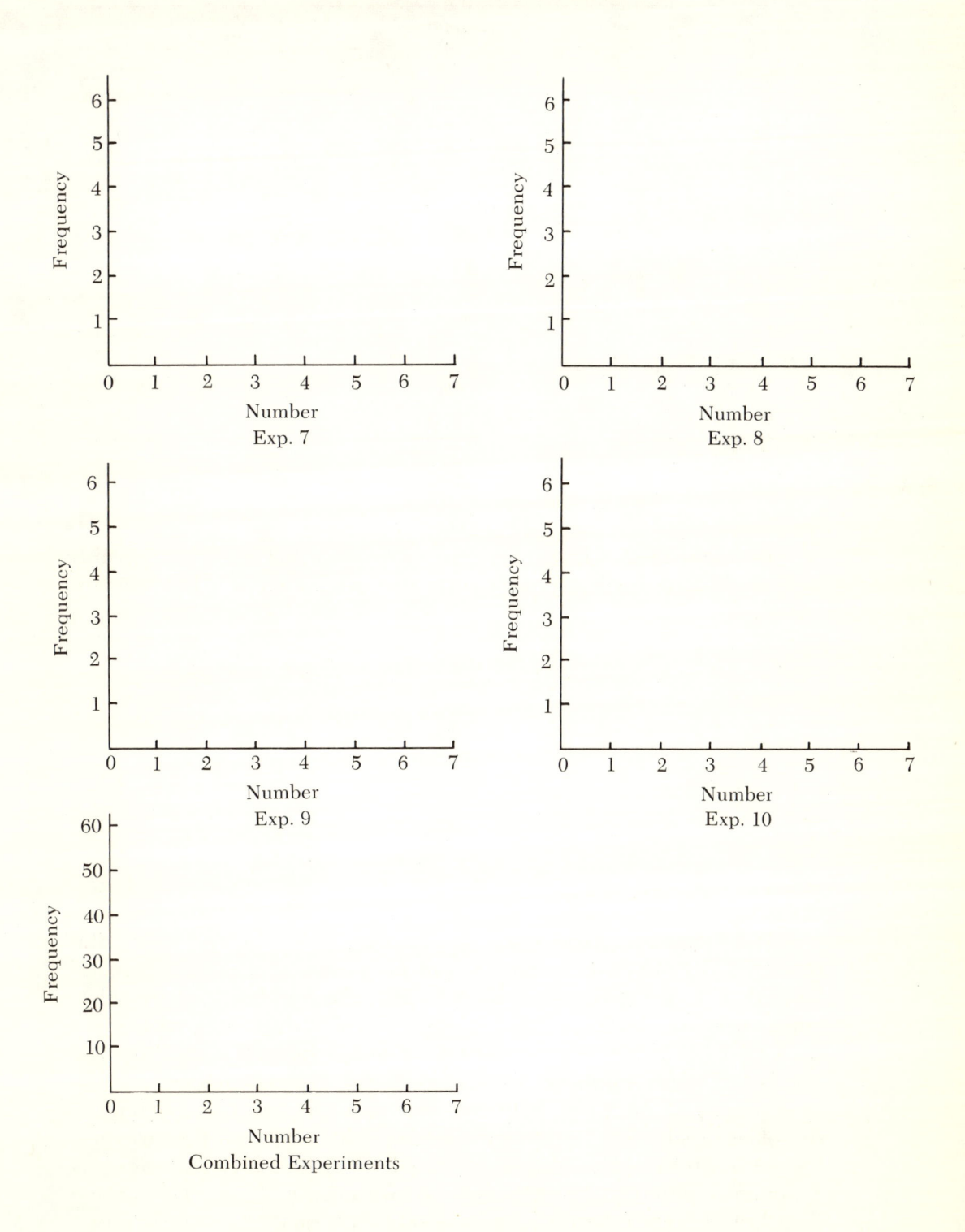

Number
Exp. 8

Number
Exp. 9

Number
Exp. 10

Number
Combined Experiments

# THE POISSON DISTRIBUTION AND ITS APPLICATIONS

**TABLE 21-1** DATA OF EXERCISE 21-1.

| Exp. No. | \multicolumn{10}{c|}{Observations} | Means |
|---|---|---|---|---|---|---|---|---|---|---|---|
| | 1 | 2 | 3 | 4 | 5 | 6 | 7 | 8 | 9 | 10 | |
| 1 | | | | | | | | | | | |
| 2 | | | | | | | | | | | |
| 3 | | | | | | | | | | | |
| 4 | | | | | | | | | | | |
| 5 | | | | | | | | | | | |
| 6 | | | | | | | | | | | |
| 7 | | | | | | | | | | | |
| 8 | | | | | | | | | | | |
| 9 | | | | | | | | | | | |
| 10 | | | | | | | | | | | |

Mean for combined experiments = _____

***Exercise 21-2:*** List at least 6 experiments that could conceivably be practical, everyday experiments that would yield a Poisson distribution.

The Poisson distribution is intimately connected with the binomial distribution. In fact, it can be shown at a more advanced level that if, in the binomial distribution, the number $N$ of trials becomes very large (100 or more) while the probability $p$ of a success for each trial becomes very small (0.01 or less)—in such a way that the product of these numbers, $N \cdot p$, is moderate in size (such as 1, or 1.2, or 0.2, or 5)—then the distribution of events takes on the characteristics of a Poisson distribution with mean equal to $Np$. For this reason—since the probability of success is small—the Poisson distribution is sometimes called the distribution of rare events. Thus, the Poisson distribution often applies in problems of quality control, when sampling for defects in an assembly line operation, so long as the probability of a defective item is very small.

## THE POISSON DISTRIBUTION AND ITS APPLICATIONS

### THE UNIFORM DISTRIBUTION IN TWO DIMENSIONS; A SPECIAL APPLICATION OF THE POISSON DISTRIBUTION

In Chapter 7 we considered the one-dimensional uniform distribution. An example was the distribution of all possible outcomes when a fair die is tossed. Imagine now a uniform distribution in two dimensions, such as the distribution of dust particles falling on a flat surface. The number of particles falling in any particular area is proportional to the size of the area if the distribution is uniform. In other words, equal-sized areas in different locations would each be expected to contain the same number of particles.

When raindrops first begin to fall, the pattern they trace on the sidewalk is an illustration of this two-dimensional uniform distribution. Also, the distribution of radioactive fallout on a given city or relatively small geographical area is uniform in two dimensions. We would not expect the raindrops or fallout to be exactly evenly spaced. Likewise, it would be very unusual to have any major concentration of them in one spot. Thus, we would expect a uniform distribution of particles in two dimensions.

A uniform distribution in either one or two dimensions is one of the most commonly observed distributions in our everyday experiences: included among the examples for the one-dimensional case are such things as outcomes when drawing straws to pay for lunch, outcomes from the toss of a die, and heads or tails when a fair coin is tossed. For the two-dimensional case, there are such examples as raindrops, snowfall, dust settling in a house, radioactive fallout, items being dropped from an aircraft (such as leaflets or items dropped by accident—bombs do not have a uniform distribution), and bird drops—particularly under a roosting tree.

If, for example, an airplane dropped 72 leaflets which fluttered to the ground in the parking area of a shopping center, and if we marked the parking area off into 36 equal-area squares, then we would expect 2 leaflets per square. However, we may observe none in some squares, one in others, two in others, and perhaps as many as six or seven in some. The numbers of leaflets landing in each square would tend to have a Poisson distribution.

*Exercise 21-3:* Name additional examples of events that you would expect to have a uniform distribution in either one or two dimensions.

# THE POISSON DISTRIBUTION AND ITS APPLICATIONS

Interestingly enough, a two-dimensional uniform distribution can be approximated by the one-dimensional Poisson distribution. We merely mark off the plane area into equal-sized squares and count the number of observations that occur in each square. This number of observations will have, approximately, a Poisson distribution.

***Exercise 21-4:*** Take a piece of paper or cardboard and mark it with 36 squares. Make six rows and six columns of squares that are about 1.5 inches on a side. Next take 36 circles of paper, each about 3/8 inch in diameter. These could be cut from a paper towel with a hole punch. Drop the paper "raindrops" one at a time, from a height of 3 or more feet above the board. They must be dropped from a height sufficient to cause them to flutter. If one lands on a line, put it in the square containing the larger part of the disk. If exactly half of it lies on either side of the line, drop it again. If a "raindrop" falls off the board, drop it again. When you are done, record the number of squares containing no drops, those containing one drop, and so forth, in Table 21-2. Notice that with 36 drops and 36 squares we would expect, on the average, one drop per square; however, we know that by sampling error there will be no drops in some squares and more than one drop in others. Figure 21-2 illustrates the Poisson distribution with mean equal to 1. If we multiply these relative frequencies by 36, we obtain the expected number of drops contained per square as given in the last column of Table 21-2 (that is, $0.369 \cdot 36 = 12.8$, which is about 13; $0.184 \cdot 36 = 6.6$, which is about 7; and so forth). Repeat the experiment two more times, for a total of three trials. Now obtain the average frequency of squares containing no drops, one drop, two drops, and so on, for the three experiments. In other words, the average for row "0" would be the sum of the frequencies for the three experiments as

**TABLE 21-2** OBSERVED DISTRIBUTION OF DROPS IN SQUARES.

| Number of Drops Contained per Square | Frequency |  |  |  | Expected Number |
|---|---|---|---|---|---|
|  | EXP. 1 | EXP. 2 | EXP. 3 | AVE. |  |
| 0 |  |  |  |  | 13 |
| 1 |  |  |  |  | 13 |
| 2 |  |  |  |  | 7 |
| 3 |  |  |  |  | 2 |
| 4 |  |  |  |  | 1 |
| 5 |  |  |  |  | 0 |
| 6 |  |  |  |  | 0 |
|  | 36 | 36 | 36 | 36 | 36 |

## THE POISSON DISTRIBUTION AND ITS APPLICATIONS

listed in that row, divided by 3; this will tell the average number of squares containing no drops in the three experiments. There will be 7 averages. Compare the results of your three experiments, and of the average of them, with the results to be expected for this experiment.

Were the averages closer to the expected numbers than were the results of the individual experiments? _____ Why or why not? _____

chapter 22

# A TEST OF FIT – THE CHI-SQUARE DISTRIBUTION

**TESTING FIT TO A SPECIFIED DISTRIBUTION**

In several experiments we have asked the question, "How much do the sample results deviate from the results expected on the basis of the model assumed?" We know that if a coin is fair, then the chances that it will land "heads" when tossed are 50 per cent. Nevertheless, if we toss the coin 50 times, even though we expect to get 25 heads, we seldom get exactly 25 heads, and if we toss it 10 times we seldom get 5 heads, because of sampling variations. From what we learned in Chapter 9, we know that if 3 coins are tossed a total of 240 times we would expect to get on the average exactly 3 heads in 30 out of these 240 tosses. We would not be surprised if in the 240 tosses we obtained 3 heads 29 times or 31 times, but if we got 3 heads 150 times, that would deviate too much from the model and would be extreme, or unusual, or surprising, or a rare event.

But just how much is "too much"? Perhaps you have been somewhat surprised at the extent to which observed results have deviated from expected results because of sampling error. The **chi-** (pronounced "ki" as in "kite") **square test**, which we shall study in this chapter, enables us to distinguish between the instances in which such deviations are due simply to sampling error and the instances in which such deviations are too large to be ascribed to sampling error. In the latter

## A TEST OF FIT—THE CHI-SQUARE DISTRIBUTION

instances, we usually conclude that the assumed model is not really appropriate. We want to be able to solve problems such as the following.

*Problem 1:* It is believed that the accident rate in a certain large foundry differs according to the day of the week. The number of accidents per weekday in this foundry is recorded for a period of one year as indicated in the following table. Do these data support the theory?

| Day of the Week | M | T | W | Th | F |
|---|---|---|---|---|---|
| Number of Accidents | 55 | 38 | 23 | 22 | 42 |

*Problem 2:* A die was tossed 72 times and "1" came up 17 times, "2" came up 10 times, and "3" came up 11 times, "4" came up 13 times, "5" came up 10 times, and "6" came up 11 times. Is the die fair? Would our answer change if the die had been tossed 720 times and the "1" came up 170 times; the "2", 100 times; the "3", 110 times; the "4", 130 times; the "5", 100 times; and the "6", 110 times? Why or why not?

*Problem 3:* A coin was tossed 10 times and heads came up 6 times. Is the coin fair? Would your answer change if the coin had been tossed 100 times and 60 heads appeared? Why or why not?

*Problem 4:* To learn whether a "buffered" aspirin tablet produced faster results than a plain aspirin, 100 people with chronic headaches were given both types on separate occasions. The patients did not know which tablet was buffered. They were asked to record the time required to produce relief. The results showed that the buffered aspirin was faster acting in 63 out of the 100 cases. Does this indicate a "true" difference in the speed of action between the two types of aspirin?

In order to apply the chi-square test to the results of an experiment, we must first determine the number of **degrees of freedom** associated with the experiment. The number of degrees of freedom, often abbreviated "*df*", is the number of classifications (or categories, or cells) that are free to vary when the total frequencies of all classifications (or categories, or cells) are held constant. In Problem 1 above, the total number of accidents is 180. Suppose someone told you that on Monday there were 55 accidents, on Tuesday 38, on Wednesday 23, and on Thursday 22. Then on Friday there were $180-55-38-23-22=42$ accidents and thus, given the total, there were only four cells that could vary freely and the number in the fifth cell was fixed by the entries in the other four.

Similarly, the number of degrees of freedom for Problem 2 is 5, since the die has 6 sides. For Problem 3 there is 1 degree of freedom, since the coin has 2 sides; and for Problem 4 there is also 1 degree of

# A TEST OF FIT—THE CHI-SQUARE DISTRIBUTION

freedom. Notice that for $k$ cells or $k$ categories of a single type, the number of degrees of freedom is $k-1$.

Another way of conceiving of degrees of freedom is the "hand-in-glove" example. As you put your first finger in a glove, you could conceivably put it in any of the five holes. The second finger could go into any of the four holes left and thus still there is a "choice" or freedom to vary. The third finger has a "choice" of three holes. The fourth finger can go into either of two holes, but the fifth finger has no choice, or no freedom to vary. Thus, for five fingers there are four degrees of freedom.

In practice we are faced with both single and multiple classification problems. If we were classifying a group of people by both hair color and eye color, it would be a double classification problem. Later in the chapter, in connection with testing fit to the model of independence, we will investigate this type of problem. The degrees of freedom in a single classification problem, such as Problems 1 through 4, are merely the number of separate classifications minus one.

It is essential to be able to distinguish between the number of trials and the number of classifications. If a die is tossed 10 times there are 6 classifications, with 5 degrees of freedom, since on *each toss* any one of 6 things can happen. If a coin is tossed 15 times there would be 1 degree of freedom, since on each toss either H or T occurs. If a coin and a die were tossed together, there would be $2 \cdot 6$ or 12 possible outcomes, which would yield 11 degrees of freedom.

Once we have determined the number of degrees of freedom associated with an experiment, we proceed to the next step in the application of the chi-square test to experimental results—the specification of the expected frequencies according to some model that we think might fit the observed results. The choice of the model with which to specify the expected frequencies depends upon the question being asked—either its positive or its negative form will usually specify some known distribution as a model.

Does the accident rate depend on the day of the week? Negation of this proposition implies that the rate has a uniform distribution over the days of the week. For Problem 1, a logical model would be to specify a uniform distribution for the five days of the week, since we were not given the specific "different" values that were believed to be true. For Problems 2 and 3, the model of a fair die or fair coin would yield a uniform distribution. For Problem 4, one would expect an equal number reporting faster action for the buffered aspirin to that for the plain aspirin if there were no true difference, and deviation from this uniform distribution would imply a true difference between the effects of the two products.

Finally, we calculate the test statistic and compare it to a tabled critical value. To obtain the test statistic, which is appropriately called "chi-square" and denoted $\chi^2$ ("$\chi$" is the Greek letter "chi"), we square the difference between the observed and the expected frequencies

## A TEST OF FIT—THE CHI-SQUARE DISTRIBUTION

for each classification or cell, and then divide the result by the expected frequency. $\chi^2$ is the sum of the quotients for all of the classifications; that is,

$$\chi^2_{df} = \sum \frac{(o-e)^2}{e},$$

where $o$ denotes the observed frequencies and $e$ denotes the expected frequencies, and the subscript on the $\chi^2$ indicates the number of degrees of freedom involved. By comparing the calculated value of $\chi^2_{df}$ with the critical values in that row of Table VIII of the appendix which corresponds to the number of degrees of freedom in our experiment, we decide whether or not the model chosen fits the observed results. The values at the top of the table are the probabilities that $\chi^2$ will be greater than or equal to the critical value in the appropriate row directly under that probability *if in fact the deviations of the observed results from the expected results are due to sampling error alone.* NOTE that for the chi-square statistic, calculated values of the test statistic that are LARGER than the tabled value are characteristic of extreme values rather than those that are smaller, as we have observed in previous tests! In other words, values of $\chi^2$ that exceed or are equal to the critical values for small probabilities indicate that the observed results are extreme, and would lead us to the conclusion that the assumed model very probably does not fit the observed results. For example, in Problem 1 referred to above, if we find by the chi-square test that the value of $\chi^2$ is extreme assuming a uniform distribution, then our sample evidence would lead us to believe that the accident rate differs according to the day of the week.

EXAMPLE 22–1. If we toss 3 coins 64 times and observe all tails 6 times, two tails and one head 20 times, two heads and one tail 30 times, and all heads 8 times, are the coins fair? The question implies the binomial model for which the expected frequencies are distributed in a 1:3:3:1 ratio. To test whether this model fits, we apply the chi-square test to the observed results as follows:

| Outcome | o | e | (o−e) | (o−e)² | $\frac{(o-e)^2}{e}$ |
|---|---|---|---|---|---|
| 3T | 6 | 8 | −2 | 4 | 0.50 |
| 2T, 1H | 20 | 24 | −4 | 16 | 0.67 |
| 2H, 1T | 30 | 24 | 6 | 36 | 1.50 |
| 3H | 8 | 8 | 0 | 0 | 0.00 |
| | $\Sigma o = 64$ | $\Sigma e = 64$ | $\Sigma(o-e) = 0$ | | $\sum \frac{(o-e)^2}{e} = 2.67$ |
| | | These should be equal (check your arithmetic) | This should be zero (to the limits of rounding error) | | |

# A TEST OF FIT—THE CHI-SQUARE DISTRIBUTION

Since we are tossing three coins, there are four possible outcomes, so the experiment has three degrees of freedom. According to Table VIII, $\chi_3^2 = 2.67$ is not extreme at all, since the table indicates that a value of 4.11 or more occurs by chance 25% of the time; thus a value of 2.67 is even less extreme.

Conclusion: based on sample evidence the coins are fair.

EXAMPLE 22-2. Suppose a sociologist is interested in whether freshmen, sophomores, juniors, or seniors tend to spend more money on recreation every week. He conducts a survey of 100 university students, of which 40 were freshmen, 30 were sophomores, 20 were juniors, and 10 were seniors. He finds the median amount spent in the entire group of 100 and then records the numbers in each class that spent more than the median amount. In the following table of results, "$o$" refers to the observed distribution. Note that the expected distribution, denoted by "$e$", is 20:15:10:5, because if there were no difference in the amount spent for recreation by classification, then the median for the whole group should also be the median for each subgroup. Thus, there would be $\frac{40}{2} = 20$ freshmen who spent more than the median amount, $\frac{30}{2} = 15$ sophomores, and so on.

| Class | $o$ | $e$ | $(o-e)$ | $(o-e)^2$ | $\frac{(o-e)^2}{e}$ |
|---|---|---|---|---|---|
| Freshmen | 32 | 20 | 12 | 144 | 7.2 |
| Sophomores | 16 | 15 | 1 | 1 | 0.1 |
| Juniors | 2 | 10 | −8 | 64 | 6.4 |
| Seniors | 0 | 5 | −5 | 25 | 5.0 |
|  | $\Sigma o = 50$ | $\Sigma e = 50$ | $\Sigma(o-e) = 0$ |  | $\Sigma \frac{(o-e)^2}{e} = 18.7$ |
|  | \multicolumn{3}{c}{Check of arithmetic} |  |  |

From the last column we see that $\chi_3^2 = 18.7$, which is larger than the value of 12.8 listed in the third row of Table VIII. Thus, the chances are less than 5 out of 1000 (or 0.5%) of getting a $\chi^2$ value this large or larger if in fact all of the classes tend to spend the same amount on recreation. The sociologist's conclusion, based on the sample data, would be that the different classes do not spend the same amount—specifically, underclassmen tend to spend considerably more than upperclassmen on recreation.

One rule that must be observed in using the chi-square distribution for this test of fit is that each expected frequency should be larger than 5. If any class has a smaller expected frequency, then we try to combine it with its neighboring classes. For example, suppose in Problem 1 the table of accidents had been:

# A TEST OF FIT—THE CHI-SQUARE DISTRIBUTION

| Day of Week | M | T | W | Th | F | Totals |
|---|---|---|---|---|---|---|
| Number of Accidents | 7 | 1 | 1 | 2 | 9 | 20 |
| Expected Number | 4 | 4 | 4 | 4 | 4 | 20 |

Since 4 is too small by the rule, one may wish to combine Monday and Friday totals and Tuesday, Wednesday, and Thursday totals, and ask instead whether the accident rate preceding and following the weekends differed from the mid-week accident rate:

| Days | M or F | T, W, or Th | Totals |
|---|---|---|---|
| Number of Accidents | 16 | 4 | 20 |
| Expected Number | 8 | 12 | 20 |

Now the expected frequencies in both categories are greater than 5, and we have reduced the degrees of freedom to one.

Now that we have examined the use of the chi-square test as a test of fit, we shall summarize briefly the steps that should be followed in applying the test:

(1) Choose the appropriate model.
(2) Make a table listing both the observed frequencies, labeled "$o$", and the expected frequencies, labeled "$e$", for each of the $k$ classifications. The totals of the frequencies in the $o$ and the $e$ columns should be the same.
(3) Ascertain the number of degrees of freedom.
(4) Be sure that each cell has $e > 5$. If not, try to combine the frequencies of some of the cells if the combination can be made in a meaningful way.
(5) Fill in a column labeled "$o - e$". Its sum should be zero, to within rounding error.
(6) Square all the items in the column of step 5 to get a column labeled $(o-e)^2$.
(7) Divide each item in the column of step 6 by the corresponding $e$ to get a column labeled $\frac{(o-e)^2}{e}$.
(8) Add the items in the column of step 7 to get $\sum \frac{(o-e)^2}{e}$.
(9) Compare the sum obtained in step 8 with the numbers in Table VIII of the appendix to find the probability of getting a value of $\chi^2_{df}$ that large or larger if in fact the specified model is correct. If the probability is very small, then the observed results are extreme and do not fit the model specified by the expected results.

# A TEST OF FIT—THE CHI-SQUARE DISTRIBUTION

**Exercise 22-1:** Answer Problem 1 at the beginning of this chapter.

**Exercise 22-2:** Answer Problem 2 at the beginning of this chapter.

**Exercise 22-3:** Answer Problem 3 at the beginning of this chapter.

**Exercise 22-4:** Answer Problem 4 at the beginning of this chapter.

# A TEST OF FIT—THE CHI-SQUARE DISTRIBUTION

**Exercise 22-5:** Test whether your sample results deviated significantly from expectation in Exercise 7-5 (page 62); in Exercises 8-9 and 8-13 (pages 68 and 69); and in Exercise 9-9 (page 79; use the total). Interpret your results.

Exercise 7-5:

Exercise 8-9:

## A TEST OF FIT—THE CHI-SQUARE DISTRIBUTION

Exercise 8-13:

Exercise 9-9:

193

# A TEST OF FIT—THE CHI-SQUARE DISTRIBUTION

**Exercise 22-6:** Test whether your sample results deviated significantly from expectation in Exercise 7-6 (page 62); in Exercises 8-10 and 8-14 (pages 68 and 69); and in Exercise 9-10 (page 81; use the total). Interpret your results.

Exercise 7-6:

Exercise 8-10:

## A TEST OF FIT—THE CHI-SQUARE DISTRIBUTION

Exercise 8-14:

Exercise 9-10:

# A TEST OF FIT—THE CHI-SQUARE DISTRIBUTION

## TESTING FIT TO THE MODEL OF INDEPENDENCE

We have considered the single classification problem at length. Suppose now that we are concerned with two classifications of items. For example, a group of people may be classified by hair color into the groups red, blonde, and brunette, and also by eye color into the groups brown, blue, green, and hazel. If all the marginal totals (in other words, the totals of all rows and columns) were known, there would be one degree of freedom for each X in the following diagram:

|  |  | Hair Color |  |  |  |
|---|---|---|---|---|---|
|  |  | R | Bl | B |  |
|  | Br | X | X |  | known |
| Eye Color | Bl | X | X |  | known |
|  | Gr | X | X |  | known |
|  | Ha |  |  |  | known |
|  |  | known | known | known |  |

The blank cells could be filled in by knowledge of the marginal totals and by knowledge of the numbers in the "X" cells. Thus, if there are $n$ items of one classification and $m$ items of a second classification in a two-way table, there will be $(n-1)(m-1)$ degrees of freedom. When we make a table as shown above for a double classification problem, we can just cover one row and cover one column, and then count the number of cells that are left to obtain the number of degrees of freedom. In this case there are $(4-1)(3-1) = 3 \cdot 2 = 6$ degrees of freedom. The simplest case, the $2 \times 2$ table, will always have only one degree of freedom. We shall be concentrating on this case for the remainder of the chapter.

The chi-square test can be used to decide whether two different classifications are or are not related to each other, as measured by data recorded in a $2 \times 2$ table. For example, we may wish to test whether a person's preference for nonpartisan candidates A or B in a political race is related to his specified political party. Suppose we denote Democrat = D and Republican = R. A special formula may be used to simplify the calculations when $\chi^2$ is used for this purpose, and, as may be easily verified, the appropriate value for the degrees of freedom will always be 1. The formula is:

$$\chi_1^2 = \frac{n\left(|ad - bc| - \frac{n}{2}\right)^2}{(a+c)(b+d)(a+b)(c+d)},$$

where the letters are the appropriate entries in the $2 \times 2$ table as follows:

## A TEST OF FIT—THE CHI-SQUARE DISTRIBUTION

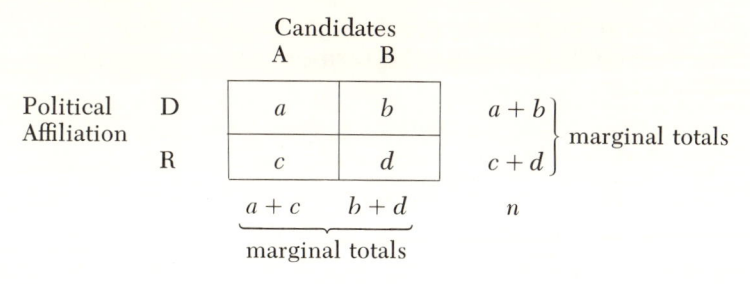

|  |  | Candidates A | B |  |
|---|---|---|---|---|
| Political Affiliation | D | $a$ | $b$ | $a+b$ ⎫ |
|  | R | $c$ | $d$ | $c+d$ ⎬ marginal totals |

$\underbrace{a+c \quad b+d}_{\text{marginal totals}}$ $\quad n$

$n$ = total frequency = $a + b + c + d$

For example, $a$ = the number of Democrats who prefer candidate A, $b$ = the number of Democrats who prefer candidate B, $c$ = the number of Republicans who prefer candidate A, $d$ = the number of Republicans who prefer candidate B, and $n$ = the total number of people surveyed. If the calculated $\chi^2$ value is less than that tabled for one degree of freedom at the desired probability level, then we would conclude that preference for candidates A or B is independent of a person's political party.

EXAMPLE 22–3. Suppose a teacher wishes to determine whether a test question discriminates between good and poor students. The teacher can find the upper 25% and the lower 25% of the students on the basis of past grades or on the basis of their total score on the test. Then he can make a 2 × 2 table listing the frequencies with which the good and poor students answered that question correctly as follows:

|  | Good Students Upper 25% | Poor Students Lower 25% |  |
|---|---|---|---|
| Question correct | 35 | 5 | 40 |
| Question wrong | 10 | 20 | 30 |
|  | 45 | 25 | 70 |

$$\chi_1^2 = \frac{70 \left( |35 \cdot 20 - 10 \cdot 5| - \frac{70}{2} \right)^2}{40 \cdot 30 \cdot 45 \cdot 25} = 19.6$$

From Table VIII we see that values of $\chi_1^2$ that are greater than or equal to 6.63 occur less than 1% of the time if in fact there is no relationship between the two factors of correct or incorrect answer and good or poor student. Thus we see that $\chi_1^2 = 19.6$ is *very* extreme if in fact no relationship exists, so we conclude that there is a relationship—in other words, the test question is related to the student's abilities and consequently is a good question to use to discriminate between good and poor students. If the $\chi^2$ value had been small, say less than 3, we would have concluded that the test question did not discriminate well between good and poor students on the basis of our sample observations.

In summary, if we perform a chi-square test on a 2 × 2 table, our

# A TEST OF FIT—THE CHI-SQUARE DISTRIBUTION

purpose is to learn whether the two classifications are related to or independent of each other. In the former case the calculated $\chi^2$ value will be equal to or larger than the critical value, and in the latter case it will be smaller than the critical value given by Table VIII. The method is as follows:

(1) Specify the observed frequencies in the form of a 2 × 2 table.
(2) Obtain all marginal totals and the overall total of all frequencies, and record them in the appropriate row and column positions of the table.
(3) Find $\chi_1^2 = \dfrac{n\left(|ad-bc|-\dfrac{n}{2}\right)^2}{(a+c)(b+d)(a+b)(c+d)}$. Notice that for this test, we do not have to obtain the expected frequencies; they are actually "built into" the formula for $\chi_1^2$.
(4) Compare the value of $\chi_1^2$ calculated from sample data to the critical values given in Table VIII for specified probabilities.
(5) Decide whether the two classifications are related ($\chi_1^2 \geq$ critical value) or are independent ($\chi_1^2 <$ critical value).

**Exercise 22-7:** Based on the following table, is there a relationship between hair color and eye color?

|  |  | Hair Color Black | Hair Color Blonde |
|---|---|---|---|
| Eye Color | Brown | 13 | 7 |
|  | Blue | 8 | 5 |

**Exercise 22-8:** Based on the following table, is there a relationship between a student's major field and his preference for problem-type or multiple choice tests?

|  |  | Major Field Mathematics | Major Field Business |
|---|---|---|---|
| Test Preference | Mult-Choice | 7 | 35 |
|  | Problems | 15 | 12 |

# A TEST OF FIT—THE CHI-SQUARE DISTRIBUTION

**Exercise 22-9:** Based on the following table, is there a relationship between an individual's sex and affiliation with the Democratic or Republican political party?

|  |  | Sex Male | Female |
|---|---|---|---|
| Political Party | Democratic | 42 | 36 |
|  | Republican | 54 | 52 |

**Exercise 22-10:** Based on the following table, is there a relationship between sex and course preference?

|  |  | Sex Male | Female |
|---|---|---|---|
| Course Preference | English | 25 | 12 |
|  | Mathematics | 48 | 24 |

# A TEST OF FIT—THE CHI-SQUARE DISTRIBUTION

### TERMS TO REMEMBER

**Chi-Square Test:** A statistical test using the chi-square statistic.

**Degrees of Freedom:** The number of categories or classifications (or cells in a table) that may vary freely if the total frequency for each classification (or marginal totals for a table) is fixed.

**Chi-Square Statistic:** $\chi^2_{df} = \sum \dfrac{(o-e)^2}{e}$ where "$o$" is the observed frequency, "$e$" is the frequency expected under the model given, and $df$ is the number of degrees of freedom involved.

chapter 23

# THE MEDIAN TEST

In Chapter 11 we investigated the sign test, and in Chapter 12 the signed ranks test, both of which enabled us to decide whether two dependent, or related, groups of observations differed in their measures of central tendency. In that case we had a design involving matched pairs or a before-after design with measures that were paired one with another. Suppose that, instead, our experimental design had two independent groups, such as we discussed in Chapter 16, and we wished to test whether the two groups have the same measure of central tendency. If the independent groups of observations have ordinal measurement scale, we may use the Mann-Whitney Test as described in Chapter 16. However, if the measurement scale is nominal, or if it is thought to be actually less precise than its ordinal or interval appearance would imply, then a median test is indicated. In summary, then, we use the median test to determine whether two independent samples of nominal measures came from populations which have the same measure of central tendency.

The procedure for the median test is as follows:

(1) Put all observations from both groups in ascending order and determine the overall median for the combined group. If nominal measure on an attribute is involved, identify in each group the total number having or not having the attribute.

(2) Make a 2 × 2 table showing a breakdown of the frequencies of observations that were above and below the median or fre-

# THE MEDIAN TEST

quencies of observations having or not having attributes as follows:

|  | Group 1 | Group 2 |  |
|---|---|---|---|
| No. of scores above median* (no. with attribute) | a | b | a + b |
| No. of scores below median* (no. without attribute) | c | d | c + d |
|  | a + c | b + d | n |

Notice that $a + c$ is the total number of observations in Group 1 and $b + d$ is the total number in Group 2. As before, this $2 \times 2$ table has one degree of freedom.

(3) Calculate $\chi_1^2$ using the same formula as in Chapter 22:

$$\chi_1^2 = \frac{n\left(|ad - bc| - \frac{n}{2}\right)^2}{(a+c)(b+d)(a+b)(c+d)}$$

(4) After calculating the $\chi^2$ value, find whether it indicates an extreme event by using Table VIII of the appendix. A value of $\chi^2$ large enough to be extreme implies that the two samples have a different median or that the two samples differ in the proportions of the attribute present, and the difference is not due to chance variation.

We pointed out in Chapter 22 that to use the special $\chi^2$ formula for a $2 \times 2$ table, we did not have to find expected frequencies. Instead they were already "built into" the formula. Here also we are taking advantage of this property of the $\chi^2$ formula for a $2 \times 2$ table. If the two groups have the same median or if they have the attribute in equal proportions, then the expected frequency for cell $a$ is the same as that for cell $b$ (and likewise, that in cell $c$ equals that in cell $d$), and the $\chi^2$ formula that we employ tests whether the observed frequencies fit this expected pattern.

In Chapter 16 we tested for equal location between two groups with the Mann-Whitney Test. It assumed ordinal measure, whereas the median test separates the data into two categories. Thus, the median test does not use as much of the sample information as the Mann-Whitney Test does, and therefore it is not so sensitive as the Mann-Whitney Test in detecting true differences. In statistical analyses, as in the marketplace, we get what we pay for. Naturally, a test that uses all of the sample information will detect significant deviations of the observed results from the assumed model more readily than will a test that ignores part of the sample data. If, however, the

---

*Those scores equal to the median can be put into either class. Try to equalize numbers in classes as much as possible.

## THE MEDIAN TEST

experimenter knows that his measurement scale lacks the precision of ordinal measurement, then in all honesty he must use a test which assumes no more precision than he actually has. The median test compares with the Mann-Whitney Test in a manner similar to the way that the sign test compares to the signed ranks test, in that it assumes less than ordinal measure.

*Exercise 23-1:* Using data gathered in Exercise 16–1, page 133, rework the problem using the median test. Does your final conclusion agree with that obtained in Chapter 16? Why or why not?

*Exercise 23-2:* Using the data gathered in Exercise 16–2, page 134, rework the problem using the median test. Does your final conclusion agree with that obtained in Chapter 16? Why or why not?

## THE MEDIAN TEST

***Exercise 23-3:*** Using the data gathered in Exercise 16-3, page 134, rework the problem using the median test. Does your final conclusion agree with that obtained in Chapter 16? Why or why not?

***Exercise 23-4:*** Using the data gathered in Exercise 16-4, page 135, rework the problem using the median test. Does your final conclusion agree with that obtained in Chapter 16? Why or why not?

chapter 24

# ESTIMATION

One of the most basic procedures of inferential statistics is **estimation:** the calculation of statistics from sample observations for the purpose of inferring from them some of the basic characteristics of the population from which the observations were drawn. When an experimenter gathers sample data, he may be interested only in describing the properties of the sample. If so, he is satisfied with the methods of descriptive statistics. On the other hand, he may intend to use the sample data to learn about the characteristics of the population from which the sample was drawn. In this case the methods of inductive or inferential statistics are required. For example, when a farmer plants two varieties of seed corn, he wishes to determine which is better so that the next year he may choose only the better variety to plant—he is not interested solely in a comparison of their bushel yields this year *except as that enables him to decide which is better for future planting.*

As we saw in Chapter 4, we can use the mean of the sample observations as an estimate of the population mean. The closeness of the sample mean to the population mean is a measure of the precision of the estimate. It depends on the size of the sample and the variability of the population. We can also use the proportion of the items of a certain type in the sample as an estimate of the proportion of items of the same type in the population. The true proportion of items of one type in the population is the theoretical probability of getting an item of that type when an item is selected at random from the population.

## ESTIMATION

The precision of proportion estimates likewise depends upon the size of the sample, as we saw when we investigated the Law of Large Numbers.

In this chapter we shall again be concerned with sample-based estimates of population means and population proportions, and we shall pay particular attention to methods of determining the accuracy of such estimates. The basic measure of precision that we shall use is the standard deviation. This time it will be the standard deviation in the distribution of the means or in the distribution of the proportions. Even as the standard deviation of a group of measurements gave us an indication of the precision of those measurements, so the standard deviations of groups of sample means and of groups of sample proportions will give us a rough idea of the accuracy of the sample means and proportions as estimates of the population means and proportions, respectively. In other words, we shall, in effect, treat the means and the proportions as if they were observations, and calculate *their* means and standard deviations.

In Exercise 21–1, we drew several samples of size 10 from a Poisson population with a mean of 2. We then used the sample means as estimates of the population mean. Our sample estimates were not always equal to 2, but often deviated from this value. We saw, how-

**TABLE 24-1** DATA FOR EXAMPLE 24-1.

| Sample Number | Observations | Means | Standard Deviations | Proportion of Ones |
|---|---|---|---|---|
| 1 | 1, 2, 3 | 2 | 1.00 | 1/3 |
| 2 | 1, 5, 6 | 4 | 2.65 | 1/3 |
| 3 | 1, 2, 6 | 3 | 2.65 | 1/3 |
| 4 | 3, 3, 3 | 3 | 0 | 0 |
| 5 | 3, 3, 4 | 3.3 | 0.58 | 0 |
| 6 | 5, 5, 6 | 5.3 | 0.58 | 0 |
| 7 | 4, 5, 5 | 4.7 | 0.58 | 0 |
| 8 | 1, 1, 2 | 1.3 | 0.58 | 2/3 |
| 9 | 2, 3, 4 | 3 | 1.00 | 0 |
| 10 | 1, 5, 6 | 4 | 2.65 | 1/3 |
| 11 | 2, 3, 4 | 3 | 1.00 | 0 |
| 12 | 2, 5, 6 | 4.3 | 2.08 | 0 |
| 13 | 2, 3, 6 | 3.7 | 2.08 | 0 |
| 14 | 2, 3, 6 | 3.7 | 2.08 | 0 |
| 15 | 2, 2, 6 | 3.3 | 2.31 | 0 |
| 16 | 3, 3, 6 | 4 | 1.73 | 0 |
| 17 | 2, 3, 6 | 3.7 | 2.12 | 0 |
| 18 | 2, 5, 6 | 4.3 | 2.08 | 0 |
| 19 | 2, 4, 5 | 3.7 | 1.53 | 0 |
| 20 | 3, 4, 5 | 4 | 1.00 | 0 |
| Averages | | 3.56 | 1.51 | 1/10 |

# ESTIMATION

ever, that the spread of the different *sample means* away from 2 was not nearly so great as the spread of the individual *sample observations* away from 2. In other words, the sample means were more accurate as estimates of the population mean than were the individual observations.

EXAMPLE 24-1. The samples of observations shown in Table 24-1 were obtained when three fair dice were tossed twenty times. For each separate experiment (each toss of three dice), both the mean and the standard deviation of the numbers on the upturned faces, and also the proportion of "ones", has been calculated. Of course, we know that the population mean in this sample space is $\mu = 3.5$, since $(1 + 2 + 3 + 4 + 5 + 6)/6 = 3.5$. Notice that the relative frequency of observations in the vicinity of $\mu = 3.5$ is far smaller than it is for the means. The average of the 20 standard deviations is 1.51, the mean based on all 60 observations is 3.56, and the average proportion of ones is approximately 1/10.

The standard deviation of the twenty means, treated as if they were twenty observations, is $s_{\bar{x}} = 0.892$. Notice that the average of the standard deviations (1.51) is approximately 1.7 (which is equal to $\sqrt{3}$ to one decimal place) times $s_{\bar{x}}$. Let us summarize the sample data in a frequency table next, to see how the relative frequencies of the observations compare with those of the means (Table 24-2).

**TABLE 24-2** FREQUENCY TABLE FOR DATA OF TABLE 24-1.

| Range | Observations | Means |
|---|---|---|
| .5 – 1.5 | 6/60 = .10 | 1/20 = .05 |
| 1.5 – 2.5 | 13/60 = .22 | 1/20 = .05 |
| 2.5 – 3.5 ⎤ * | 14/60 = .23 ⎤ * | 6/20 = .30 ⎤ * |
| 3.5 – 4.5 ⎦ | 6/60 = .10 ⎦ | 10/20 = .50 ⎦ |
| 4.5 – 5.5 | 10/60 = .17 | 2/20 = .10 |
| 5.5 – 6.5 | 11/60 = .18 | 0/20 = .00 |
|  | 1.00 | 1.00 |

*80% of the means are within one unit of 3.5, but only 33% of the observations are within one unit of 3.5.

**Exercise 24-1:** (a) Calculate the standard deviations of your first five samples drawn in Exercise 21-1 (page 176). Then calculate the average of these five standard deviations. (Treat the 5 standard deviations as if they were just 5 observations, and calculate their mean.)

ESTIMATION

(b) Calculate the standard deviation of the means of the first five samples drawn in Exercise 21-1. (In other words, treat these 5 means as if they were just 5 observations, and calculate the standard deviation.)

(c) Compare the results obtained in (a) with that of (b). Which was smaller? _____ How much smaller? _____

_____

**Exercise 24-2:** (a) Calculate the standard deviations of your last five samples drawn in Exercise 21-1. Then calculate the average of these five standard deviations. (Treat the 5 standard deviations as if they were just 5 observations, and calculate their mean.)

(b) Calculate the standard deviation of the means of the last five samples drawn in Exercise 21-1. (In other words, treat these 5 means as if they were just 5 observations, and calculate the standard deviation.)

(c) Compare the results obtained in (a) with that of (b). Which was smaller? _____ How much smaller? _____

In working with your samples in Exercises 24-1 and 24-2 you should have noticed that the average standard deviation of the samples was larger than the standard deviations of the means of the samples. Thus, you should be able to observe from your own sample data the fact that the means are dispersed far less than are the observations themselves. With more sampling and calculations, we could also observe that, if the size of the samples is increased, then the dispersion

# ESTIMATION

of the sample means will decrease. This agrees with our intuition because we expect a larger sample to give us a better measure of the true point of central tendency than would a smaller sample. Thus, the degree of precision seems to be intimately related to the number of observations in the sample. Now we shall see exactly what that relationship is.

It is known in statistical theory (see Chapter 26 for more details on the Central Limit Theorem) that, under very general conditions, the standard deviation of the means of several samples drawn from the same population can be approximated by the quotient of the standard deviation of a single sample and the square root of the sample size $n$; in other words, by $\frac{s}{\sqrt{n}}$. This quantity $\frac{s}{\sqrt{n}}$ is usually denoted $s_{\bar{x}}$, and is called the **standard error,** or standard deviation of the sample mean. The standard error is an estimate of the standard deviation of all the means of samples of size $n$ in the population. Notice that its denominator is $\sqrt{n}$, so that as the sample size increases, $s_{\bar{x}}$ decreases, indicating—as we suspected—that a larger sample size increases the precision with which we can estimate the population mean by using $\bar{x}$.

Consequently, $s_{\bar{x}}$ may be considered a measure of the degree of precision with which the population has been estimated by the sample mean. If, in a set of sample data, $s_{\bar{x}}$ is small relative to the size of $\bar{x}$, then the population mean has been estimated with a high degree of precision. If $s_{\bar{x}}$ is large relative to the size of the sample mean, then there is not much precision in the estimation of the population mean.

EXAMPLE 24-2. For the data of the twenty samples from Example 24-1, the values of $s_{\bar{x}}$ for each sample would be as shown in Table 24-3.

Just as we used the standard error to get an idea of the accuracy of our estimates of the population mean, we can also obtain an indication of the accuracy of our estimates of population proportions by using the quantity $s_p = \sqrt{\frac{p(1-p)}{n}}$, where $p$ is the sample proportion and $n$ is the sample size. This quantity $s_p$ is an estimate of the standard deviation of the proportions in all samples of size $n$ that could be taken from the population, and is often called the **standard error of the proportion.** Thus, our estimates of $p$ will be considered precise whenever $s_p$ is small relative to $p$.

EXAMPLE 24-3. If we toss a coin 100 times and observe 45 heads, then our sample estimate of $p$, the chance of getting a "head" with this coin, is $p = \frac{45}{100} = 0.45$, and $s_p = \sqrt{\frac{(0.45)(0.55)}{100}} = 0.0497$. As we can see, $s_p$ is rather small compared to $p$.

EXAMPLE 24-4. For the data of Example 24-1, the values of $s_p$ for each individual sample of size three are listed in Table 24-4.

**TABLE 24-3** CALCULATION OF STANDARD ERROR.

| Sample Number | Mean | s | $s_{\bar{x}} = \dfrac{s}{\sqrt{3}}$ |
|---|---|---|---|
| 1 | 2 | 1.00 | 0.577 |
| 2 | 4 | 2.65 | 1.530 |
| 3 | 3 | 2.65 | 1.530 |
| 4 | 3 | 0 | 0 |
| 5 | 3.3 | 0.58 | 0.335 |
| 6 | 5.3 | 0.58 | 0.335 |
| 7 | 4.7 | 0.58 | 0.335 |
| 8 | 1.3 | 0.58 | 0.335 |
| 9 | 3 | 1.00 | 0.577 |
| 10 | 4 | 2.65 | 1.530 |
| 11 | 3 | 1.00 | 0.577 |
| 12 | 4.3 | 2.08 | 1.201 |
| 13 | 3.7 | 2.08 | 1.201 |
| 14 | 3.7 | 2.08 | 1.201 |
| 15 | 3.3 | 2.31 | 1.334 |
| 16 | 4 | 1.73 | 0.999 |
| 17 | 3.7 | 2.12 | 1.224 |
| 18 | 4.3 | 2.08 | 1.201 |
| 19 | 3.7 | 1.53 | 0.883 |
| 20 | 4 | 1.00 | 0.577 |

$$\bar{s} = \frac{\Sigma s}{20} = 1.51 \qquad \bar{s}_{\bar{x}} = \frac{\Sigma s_{\bar{x}}}{0} = 0.874$$

$$\frac{\bar{s}}{\sqrt{n}} = 0.872$$

$s_{\bar{x}}$ calculated using means (see Example 24-1) = 0.892.

## ESTIMATION

**TABLE 24-4** CALCULATION OF $s_p$.

| Sample Number | Proportions (p) of Ones | $s_p = \sqrt{\dfrac{p(1-p)}{n}}$ |
|---|---|---|
| 1 | 1/3 | 0.272 |
| 2 | 1/3 | 0.272 |
| 3 | 1/3 | 0.272 |
| 4 | 0 | 0 |
| 5 | 0 | 0 |
| 6 | 0 | 0 |
| 7 | 0 | 0 |
| 8 | 2/3 | 0.272 |
| 9 | 0 | 0 |
| 10 | 1/3 | 0.272 |
| 11 | 0 | 0 |
| 12 | 0 | 0 |
| 13 | 0 | 0 |
| 14 | 0 | 0 |
| 15 | 0 | 0 |
| 16 | 0 | 0 |
| 17 | 0 | 0 |
| 18 | 0 | 0 |
| 19 | 0 | 0 |
| 20 | 0 | 0 |

$s_p = \sqrt{\dfrac{\bar{p}(1-\bar{p})}{60}} = 0.039$      $\bar{p} = 0.10$      $\bar{s}_p = 0.068$

(Based on "one sample" of size 60)      (Based on samples of size 3)

Compare them with the value of $s_p$ calculated by grouping the data into one sample of size 60.

***Exercise 24-3:*** Using a sampling urn and ignoring all colors except "white" and "non-white," estimate the proportion of white balls in the urn on the basis of samples of sizes 1, 10, and 25. Do this 10 times for each sample size, and estimate $p =$ the chance of getting a white ball on the basis of the mean value of $p$ taken in all ten samples of size 25; fill in the following table with your results.

# ESTIMATION

| Sample Number | Estimates of p Based on Sample Sizes |    |    |
|---|---|---|---|
|   | 1 | 10 | 25 |
| 1 |   |   |   |
| 2 |   |   |   |
| 3 |   |   |   |
| 4 |   |   |   |
| 5 |   |   |   |
| 6 |   |   |   |
| 7 |   |   |   |
| 8 |   |   |   |
| 9 |   |   |   |
| 10 |  |   |   |

Value of $p = P$(white) based on 10 samples of 25 each _____

**Exercise 24-4:** Using dice and recording only "one" or "not-one," estimate the proportion of ones on the basis of throwing 1, 10, and 25 dice ten times each. Next estimate $p =$ the chance of getting a "one" by obtaining the average of all the $p$'s calculated for the ten samples of size 25; fill in the following table with your results.

| Sample Number | Estimates of p Based on Sample Sizes |    |    |
|---|---|---|---|
|   | 1 | 10 | 25 |
| 1 |   |   |   |
| 2 |   |   |   |
| 3 |   |   |   |
| 4 |   |   |   |
| 5 |   |   |   |
| 6 |   |   |   |
| 7 |   |   |   |
| 8 |   |   |   |
| 9 |   |   |   |
| 10 |  |   |   |

Value of $p = P$(one) based on 10 samples of 25 each _____

**Exercise 24-5:** Throw 10 thumbtacks on the floor (no rug) or on a large tabletop and count the number that land point up. Repeat 49 times and, on the basis of your sample data, fill in the following table.

**ESTIMATION**

| Trial | Number That Landed Point Up | Sample Estimate of p | Sample Estimate of $s_p$ |
|---|---|---|---|
| 1 | | | |
| 2 | | | |
| 3 | | | |
| 4 | | | |
| 5 | | | |
| 6 | | | |
| 7 | | | |
| 8 | | | |
| 9 | | | |
| 10 | | | |
| 11 | | | |
| 12 | | | |
| 13 | | | |
| 14 | | | |
| 15 | | | |
| 16 | | | |
| 17 | | | |
| 18 | | | |
| 19 | | | |
| 20 | | | |
| 21 | | | |
| 22 | | | |
| 23 | | | |
| 24 | | | |
| 25 | | | |

# ESTIMATION

| Trial | Number That Landed Point Up | Sample Estimate of p | Sample Estimate of $s_p$ |
|---|---|---|---|
| 26 | | | |
| 27 | | | |
| 28 | | | |
| 29 | | | |
| 30 | | | |
| 31 | | | |
| 32 | | | |
| 33 | | | |
| 34 | | | |
| 35 | | | |
| 36 | | | |
| 37 | | | |
| 38 | | | |
| 39 | | | |
| 40 | | | |
| 41 | | | |
| 42 | | | |
| 43 | | | |
| 44 | | | |
| 45 | | | |
| 46 | | | |
| 47 | | | |
| 48 | | | |
| 49 | | | |
| 50 | | | |

## ESTIMATION

1. What is the range of sample probabilities that you calculated?

   _____

2. Compute $\bar{p} = \dfrac{\Sigma p}{50} =$ _____

3. Compute $\bar{p}$ by taking the over-all sum of the "number that landed point up" column and dividing by 500; $\bar{p} =$ _____

   Does this agree with the answer in part 2? _____

   Why or why not? _____

   _____

4. Based on your sample data, what would you estimate as the true value of $p$? $p =$ _____

5. Compute the standard deviation of your sample estimates of $p$ using the formula $s_p = \sqrt{\dfrac{\Sigma p^2 - \dfrac{(\Sigma p)^2}{50}}{49}} =$ _____

6. How does $s_p$ calculated in part 5 compare with the mean of the individual $s_p$ values calculated in each sample? _____

   _____

**Exercise 24–6:** Throw 10 dice and count how many show an even number on their upturned faces. Repeat 49 times and, on the basis of your sample data, fill in the following table.

| Trial | Number of Even Faces | Sample Estimate of p | Sample Estimate of $s_p$ |
|---|---|---|---|
| 1 | | | |
| 2 | | | |
| 3 | | | |
| 4 | | | |
| 5 | | | |
| 6 | | | |
| 7 | | | |
| 8 | | | |
| 9 | | | |
| 10 | | | |
| 11 | | | |

**ESTIMATION**

| Trial | Number of Even Faces | Sample Estimate of p | Sample Estimate of $s_p$ |
|---|---|---|---|
| 12 | | | |
| 13 | | | |
| 14 | | | |
| 15 | | | |
| 16 | | | |
| 17 | | | |
| 18 | | | |
| 19 | | | |
| 20 | | | |
| 21 | | | |
| 22 | | | |
| 23 | | | |
| 24 | | | |
| 25 | | | |
| 26 | | | |
| 27 | | | |
| 28 | | | |
| 29 | | | |
| 30 | | | |
| 31 | | | |
| 32 | | | |
| 33 | | | |
| 34 | | | |
| 35 | | | |
| 36 | | | |
| 37 | | | |
| 38 | | | |
| 39 | | | |
| 40 | | | |
| 41 | | | |
| 42 | | | |
| 43 | | | |
| 44 | | | |
| 45 | | | |
| 46 | | | |
| 47 | | | |
| 48 | | | |
| 49 | | | |
| 50 | | | |

# ESTIMATION

1. What is the range of sample probabilities that you calculated?

   _____

2. Compute $\bar{p} = \dfrac{\Sigma p}{50} =$ _____

3. Compute $\bar{p}$ by taking the over-all sum of the "number of even faces" column and dividing by 500; $\bar{p} =$ _____

   Does this agree with the answer in part 2? _____

   Why or why not? _____

   _____

4. Based on your sample data, what would you estimate as the true value of $p$? $p =$ _____

5. Compute the standard deviation of your sample estimates of $p$ using the formula $s_p = \sqrt{\dfrac{\Sigma p^2 - \dfrac{(\Sigma p)^2}{50}}{49}} =$ _____

6. How does $s_p$ calculated in part 5 compare with the mean of the individual $s_p$ values calculated in each sample? _____

   _____

Each individual observation has a considerably greater tendency to spread or fall away from the population mean than does an average measurement. That is to say, the sample averages tend to cluster much more closely about the overall average or true population value than do the individual measurements. Likewise, a proportion may be estimated much more precisely based on a large sample than on a small sample. These facts are obvious when one notices that both $s_{\bar{x}}$ and $s_p$ have $\sqrt{n}$ in their denominators, so that as $n$ gets large, $s_{\bar{x}}$ and $s_p$ both get smaller.

*Brain teaser:* At this point you are "on your own." Obtain 5 sampling urns, labeled A, B, C, D, and E, each containing at least one black ball and at least one white ball. Let your instructor put the black and white balls in so that you have no knowledge of how many of each color are in each urn. It makes no difference how many balls are contained in each urn. Your task is to estimate the proportion of black balls in each urn. Design your experiment, conduct your experiment, analyze your data, and summarize your results. Be clear and explicit in *explaining what you did, how you did it, what your results were, and how precise your estimates are.*

**TERMS TO REMEMBER**

**Estimation:** The calculation of statistics from sample observations for the purpose of inferring from them some of the basic characteristics of the population from which the observations were drawn.

**Standard Error:** Standard deviation of the mean. $s_{\bar{x}} = \frac{s}{\sqrt{n}}$, where $s$ is the sample standard deviation and $n$ is the sample size.

**Standard Error of a Proportion:** $s_p = \sqrt{\frac{p(1-p)}{n}}$ = the standard deviation of the distribution of sample proportions taken from samples of size $n$.

chapter 25

# DECISION MAKING

Statistics has been called "the art of decision making under uncertainty." We are constantly called upon to make decisions. The businessman must decide whether to buy or sell stocks, whether to build a new manufacturing plant or open a new retail outlet, when and from whom to buy raw materials, or whether to hire an applicant for an existing job vacancy. The housewife must decide what to serve her family for dinner, when to do her grocery shopping, and what brand of appliance to buy. The student must decide whether to take a certain course this term or next term, what electives to schedule, or whether to cut that English class in order to get an earlier ride home this week-end. The thoughtful person tries to make his decisions wisely, making use of all available information. As a matter of fact, gathering data and analyzing it prior to making a decision is merely an extension of our own everyday experience. Whether or not we apply some specific technique, our daily decisions are made using the same type of reasoning that is used by the statistician.

**Exercise 25-1:** To illustrate the statistical methods used in one type of problem, let us suppose that we have an urn with some black and some white balls in it. By means of a sampling procedure we want to decide whether the urn has a certain proportion of white balls in it. Suppose we think that 1/10 of the balls in the urn are white and want to test whether this is the true proportion. If you drew one ball at a

# DECISION MAKING

time, observed its color and then replaced it before drawing the next ball, how many balls would you expect to have to draw from the urn before you would observe one white ball? _____ Would you be surprised if no white balls were observed in 11 trials? _____ In 12 trials? _____ In 20 trials? _____ Would the proportion of white balls change if you did not replace the balls drawn? _____

**Exercise 25-2:** Suppose you have a bag of jellybeans which you believe has only 1/15 black jellybeans, and by some sampling procedure you want to decide whether this is the true proportion. Suppose the black ones are your favorite, but you cannot see what color you are getting when you draw from the bag. How many jellybeans would you expect to have to draw in order to get just one black one? _____ Would you be surprised if there were no black ones in the first 16 jellybeans drawn? _____ In the first 17? _____ In the first 35? _____ If you ate the black ones as you drew them out, would this change the proportion of black jellybeans in the bag?

---

How many trials would be necessary to decide whether the proportion is really 1/10 in Exercise 25-1 or 1/15 in Exercise 25-2? This question is a bit difficult to answer. In more advanced courses we find that the sample size $n$ needed in order to be about 95% sure that we have kept the error within 0.01 of the true value of the proportion $p$ is given approximately by $40,000\, p(1-p)$. Thus, to test the hypothesis that $p = 1/10$, since $40,000\,(0.1 \times 0.9) = 3600$, it would be necessary to take at least 3600 observations. To test $p = 1/15$, it would be necessary to take at least 10,000 observations. That could be tedious! If we are willing to accept an error of 0.1 rather than 0.01, then we need about $400\, p(1-p)$ observations. "An error of 0.1," for example, means that when we estimate $p = 0.5$, we are 95% confident that $p$ lies between 0.4 and 0.6. By "an error of 0.01," we mean that we are 95% confident that $p$ lies between 0.49 and 0.51.

**Exercise 25-3:** Decide how many times a coin should be tossed to decide whether it is fair. _____ Test a coin and decide whether it is fair. How many heads did you get? _____ How

## DECISION MAKING

many tails did you get? _____ Is the coin fair? _____

What size error did you specify? _____

**Exercise 25-4:** Decide how many times a die should be tossed to find out whether there is a fair proportion of even numbers.

_____

Test a die and decide whether the proportion of even numbers is fair.

How many even numbers did you get? _____ How many odd

numbers did you get? _____ What size error did you specify?

_____ What else do you suspect you would have to do to show

whether the die is fair? _____

**Exercise 25-5:** Set up a sampling bag with jellybeans or gumdrops in which there is a 1/3 chance of drawing some particular color, such as red. Assuming that you want to be 95% sure of getting an error of no more than 0.1, decide by sampling whether the proportion of that color is 1/3. (Note: don't eat the gumdrops or jellybeans as you draw them out or the probabilities will change!)

**Exercise 25-6:** Set up a sampling urn with 1/4 chance of getting a white ball. Assuming that you want to be 95% sure of getting no more than an error of 0.1, decide by sampling whether the proportion of white balls is 1/4. (Note: Be sure you replace after each draw or the probabilities will change!)

## DECISION MAKING

This decision making aspect of statistics is really nothing new for us. We have already employed it in many statistical tests. In such tests as the sign test, the signed ranks test, the median test, and the Mann-Whitney $U$ Test, we gathered sample data and calculated a test statistic; and then on the basis of a tabled critical value, we decided whether or not the measures of central tendency of two samples were alike or different. We chose our test statistic from the table according to the probability of error we were willing to accept in making this decision.

In Chapter 6 we used a test statistic to decide whether sample observations were in a random sequence, and we used the Wald-Wolfowitz Runs Test of Chapter 20 to decide whether two samples came from populations that differed in one or a combination of the four basic properties. Using the chi-square test we were able to decide, on the basis of a calculated test statistic, whether a sample of observations could be assumed to fit a given model. As we review the various techniques we have investigated, it becomes obvious why statistics is called the art of decision making under uncertainty.

chapter 26

# THE NORMAL DISTRIBUTION

In Chapter 9 we considered the binomial distribution. Since then we have used it upon several occasions. Notice in Figures 26-1, 26-2, and 26-3 how the shape of the binomial distribution changes as $n$ becomes large. We have assumed $p = 0.5$ in all three cases for convenience. From these figures we see that as $n$ becomes large, the binomial distribution begins to resemble more closely a distribution which is called the **normal distribution.**

One of the most important distributions used in statistical theory and application is this normal distribution. Its mathematical model is a bell-shaped smooth curve which extends infinitely far in both positive and negative directions above the real axis, is centered at its mean $\mu$, and has standard deviation of $\sigma$, as shown in Figure 26-4. Notice that the curve is symmetric with respect to its mean and is mesokurtic, as was pointed out in Chapter 19.

One of the characteristic properties of this normal distribution also gives us a way to visualize the true meaning of the standard deviation as a measure of spread. Notice in Figure 26-4 that in the normal distribution, one standard deviation unit (one $\sigma$-unit) to either side of the mean is the point at which this bell-shaped symmetric curve changes its curvature from concave to convex—in other words, instead of being "cupped downward," at that particular distance from the mean it begins to "cup upward."

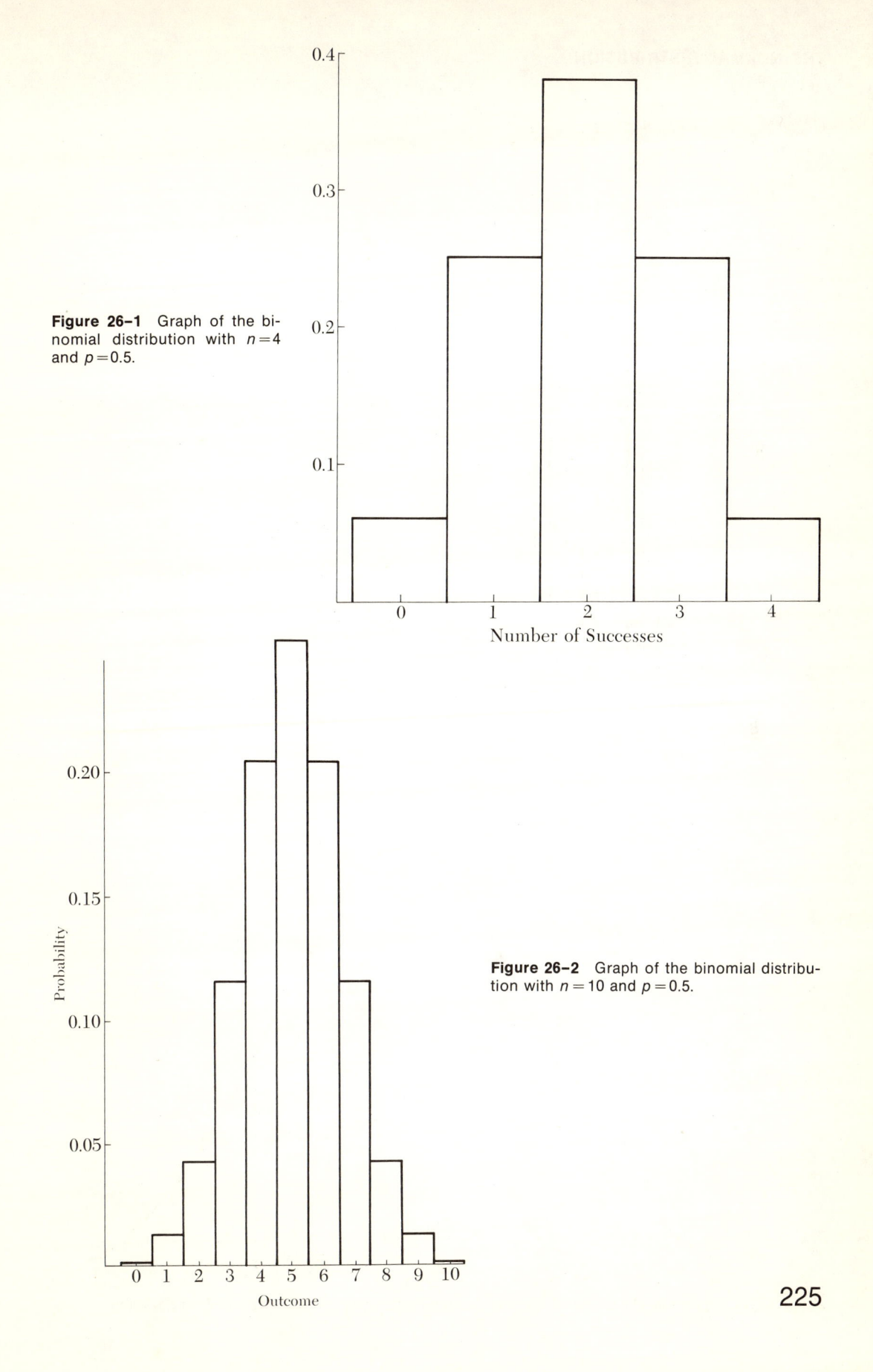

**Figure 26-1** Graph of the binomial distribution with $n=4$ and $p=0.5$.

**Figure 26-2** Graph of the binomial distribution with $n=10$ and $p=0.5$.

## THE NORMAL DISTRIBUTION

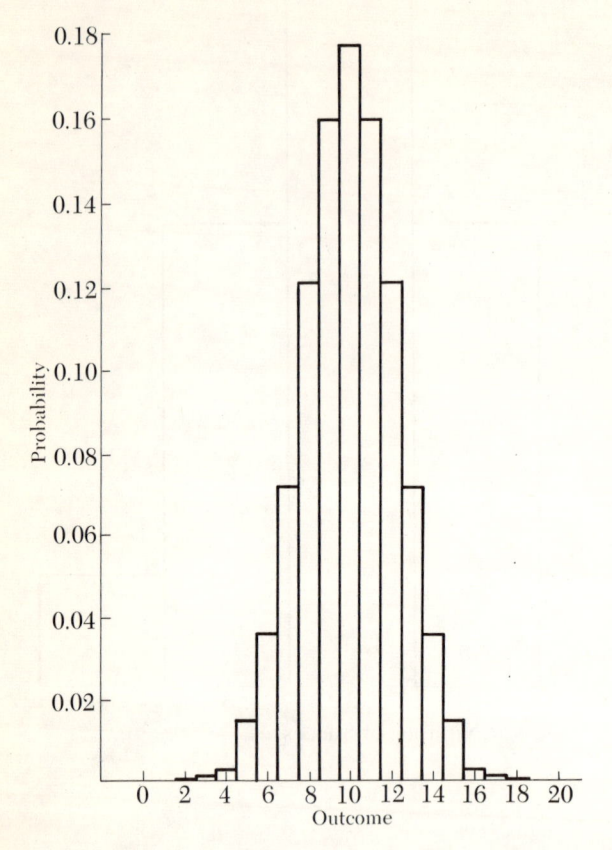

**Figure 26-3** Graph of the binomial distribution with $n = 20$ and $p = 0.5$.

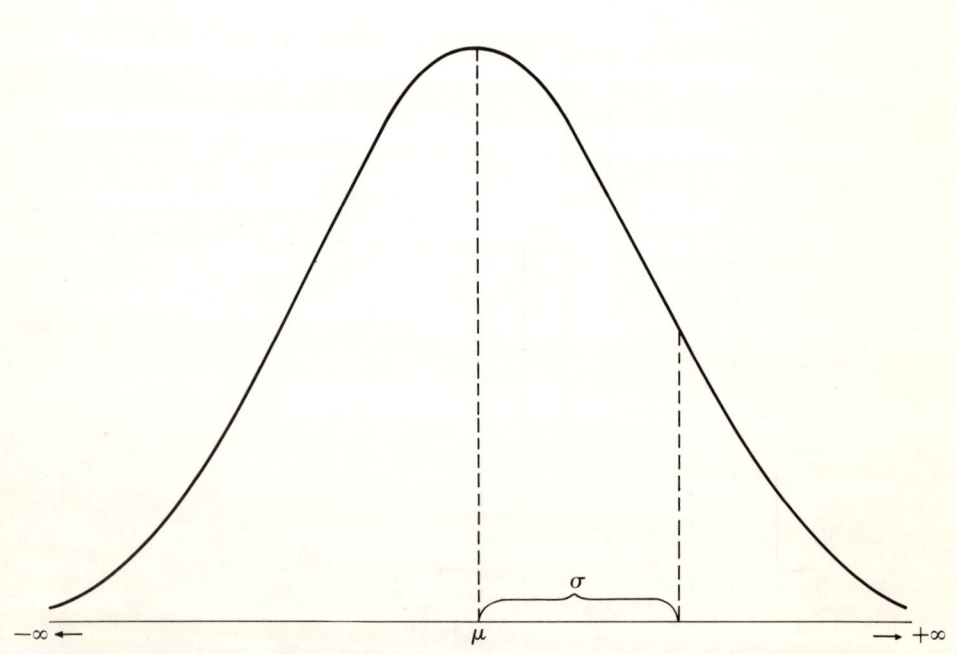

**Figure 26-4** Graph of the normal distribution with mean $\mu$ and standard deviation $\sigma$.

# THE NORMAL DISTRIBUTION

There are many everyday measurements that are normally distributed. Among these are such things as measurement errors, IQ's, scores of children on achievement tests, heights of corn stalks, and people's heights. Many biological, psychological, and sociological variables are normally distributed.

One of the most important uses of the normal distribution results from the **Central Limit Theorem.** This theorem asserts that if we take random samples from any population (with any distribution!) and observe the distribution of the means of all samples of size $n$ that could be drawn from this general population, then if $n$ is large enough (30 or more), the distribution of the sample means will look very much like a normal distribution. Its mean will be equal to that of the original population and its standard deviation will be equal to the original standard deviation divided by the square root of the sample size.

This Central Limit Theorem is illustrated by Figure 26–5. Notice in particular in Figure 26–5 that the populations sampled are quite different. Population 1 has a uniform distribution. Population 2 is V-shaped. Population 3 looks like a reversed J, while population 4 has a normal distribution. In spite of these differences, when the graph of the distribution of the means of all possible samples of size 30 is drawn, it has the bell-shaped, symmetric characteristics of the normal distribution. Notice also that the means of the distribution of means are equal to the means of the original populations. Even in extremely small samples, like size 2, the distribution of the means is already tending toward the normal shape. In particular, the least normal population—population 2—already has the beginnings of a dominant mode, or high point, for $n = 2$. In population 4 especially we may observe the effect of sample size upon the dispersion of the distribution.

Recall that in Chapter 24 we learned that for samples of size $n$, the standard deviation of the means was equal to the sample standard deviation divided by the square root of the sample size, or $s_{\bar{x}} = \dfrac{s}{\sqrt{n}}$.

We remarked at that time that the sample size affected the precision with which the mean could be measured. In Figure 26–5, we see that the spread of the distribution away from its point of central tendency is cut somewhat for $n = 2$, is more than halved for $n = 5$, and is drastically reduced for $n = 30$. For all populations except population 3, the distributions originally are symmetric and, of course, the distributions of the means remain symmetric for all sample sizes. For population 3, the original distribution is extremely right-skewed; still, even though some effects of skewness are apparent in small samples, in samples of size 30 the skewness has disappeared and the characteristic symmetry and normal shape is apparent.

Often in practice the experimenter wishes to draw conclusions about the means of underlying populations based on samples from these populations. This knowledge that the means will tend to be

## THE NORMAL DISTRIBUTION

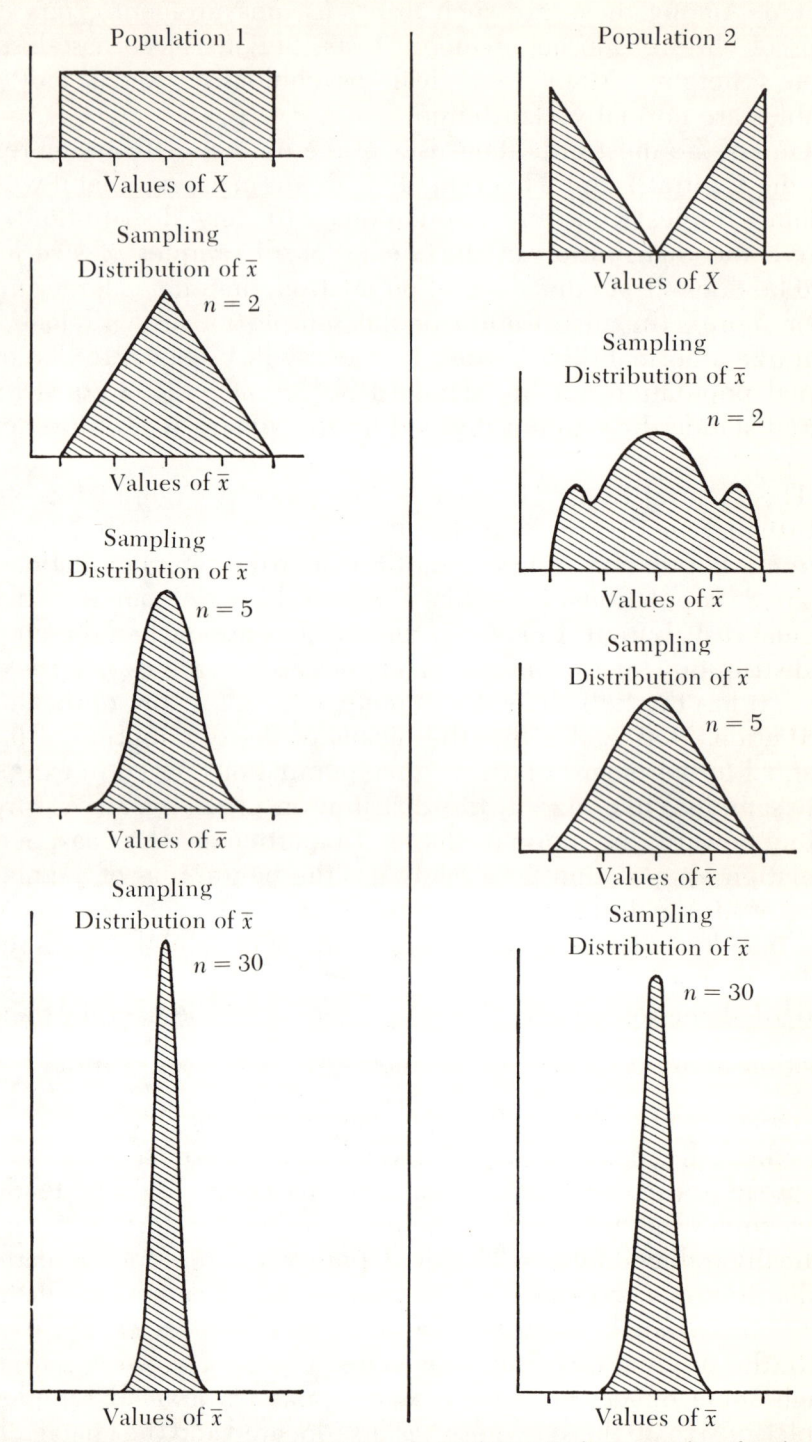

**Figure 26–5** Illustration of the distribution of sample means from samples of various sizes taken from different basic populations.

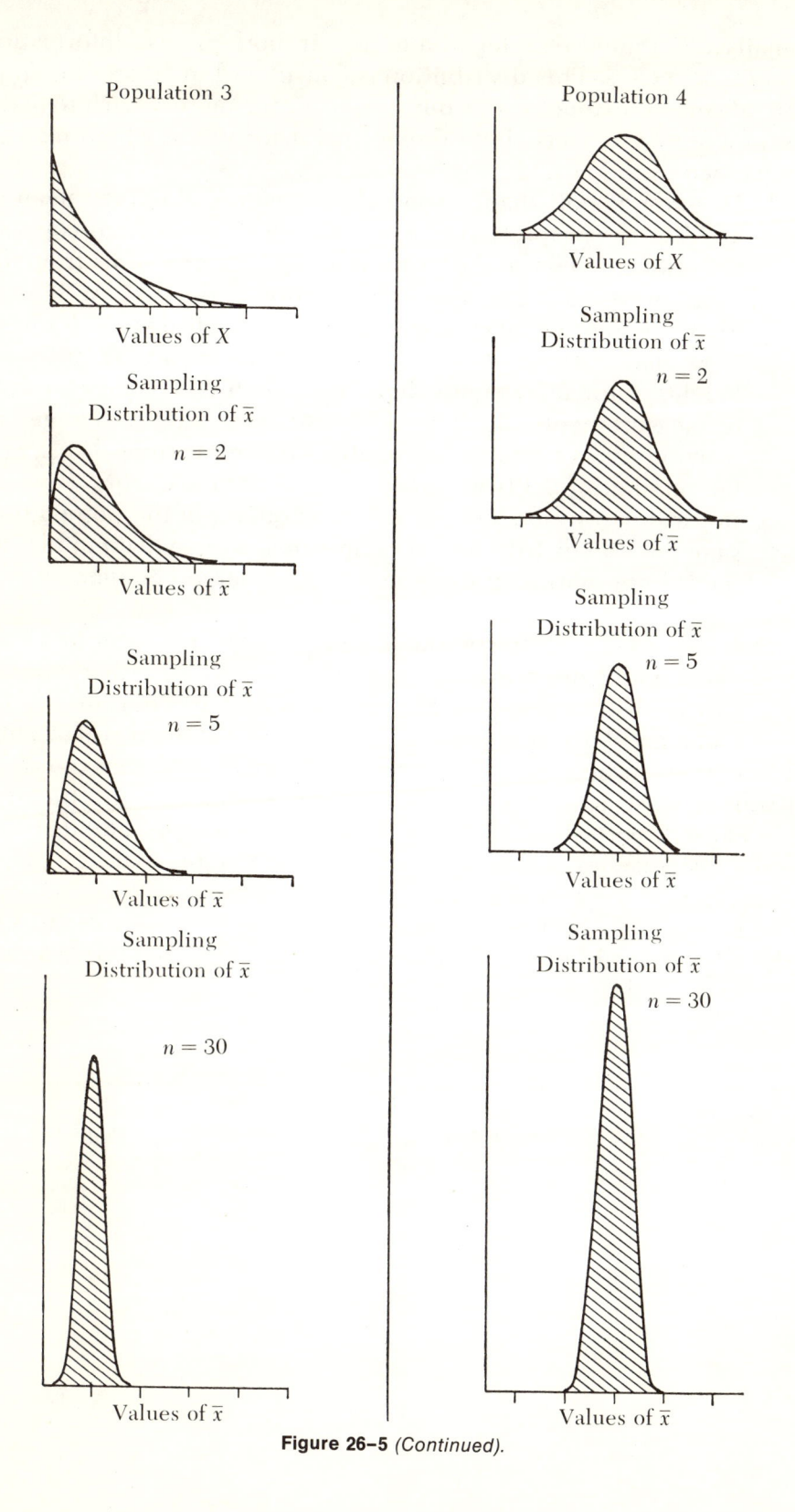

# THE NORMAL DISTRIBUTION

Population 3

Values of X

Sampling Distribution of $\bar{x}$
$n = 2$

Values of $\bar{x}$

Sampling Distribution of $\bar{x}$
$n = 5$

Values of $\bar{x}$

Sampling Distribution of $\bar{x}$
$n = 30$

Values of $\bar{x}$

Population 4

Values of X

Sampling Distribution of $\bar{x}$
$n = 2$

Values of $\bar{x}$

Sampling Distribution of $\bar{x}$
$n = 5$

Values of $\bar{x}$

Sampling Distribution of $\bar{x}$
$n = 30$

Values of $\bar{x}$

**Figure 26-5** *(Continued).*

## THE NORMAL DISTRIBUTION

normally distributed enables him to get far more precise information about these means. This distribution is considered in greater detail in more advanced textbooks. For our purposes, we merely wish to make note of a few basic facts about the normal distribution which may be summarized as follows:

1. Its graph is bell-shaped, smooth, symmetric about its mean $\mu$, and has the property that its curvature changes from concave downward to concave upward at a distance of $\sigma$ units from the mean, where $\sigma$ is the standard deviation of the distribution.
2. Many biological, psychological, and sociological variables have a normal distribution, including such things as people's heights; animals' weights, heights, and growth changes; errors of measurements; achievement test scores; IQ's; changes in reaction times for rats after treatments with various drugs; and the number of flat tires per day in a fleet of taxicabs.
3. Regardless of the shape of the distribution of the populations sampled, the distribution of sample means from that population tends to be normal if the sample size is large enough.

*Exercise 26-1:* Using a small wedge (like a doorstop) or inclined plane and some disks, one may take observations that will follow a normal distribution. Set the disk at a certain point on the inclined plane and let it roll down the plane onto a smooth surface. Try it a few times to be sure that the starting point is such that the disks will roll at least a few inches away from the bottom of the inclined plane but will not roll off the desk or table top. Measure the distance that the disk rolls away from the inclined plane to the nearest tenth of an inch. Repeat the experiment, always starting the disk from the same point. Record your measurements until you have 50 observations. Fill in the following table with the observations and then put the raw data points in ascending numerical order.

| Number of Trial | Observations | Ordered Observations | Frequency |
|---|---|---|---|
| 1 | | | |
| 2 | | | |
| 3 | | | |
| 4 | | | |
| 5 | | | |
| 6 | | | |
| 7 | | | |
| 8 | | | |
| 9 | | | |

## THE NORMAL DISTRIBUTION

| Number of Trial | Observations | Ordered Observations | Frequency |
|---|---|---|---|
| 10 | | | |
| 11 | | | |
| 12 | | | |
| 13 | | | |
| 14 | | | |
| 15 | | | |
| 16 | | | |
| 17 | | | |
| 18 | | | |
| 19 | | | |
| 20 | | | |
| 21 | | | |
| 22 | | | |
| 23 | | | |
| 24 | | | |
| 25 | | | |
| 26 | | | |
| 27 | | | |
| 28 | | | |
| 29 | | | |
| 30 | | | |
| 31 | | | |
| 32 | | | |
| 33 | | | |
| 34 | | | |
| 35 | | | |
| 36 | | | |
| 37 | | | |
| 38 | | | |
| 39 | | | |
| 40 | | | |
| 41 | | | |
| 42 | | | |
| 43 | | | |
| 44 | | | |

## THE NORMAL DISTRIBUTION

| Number of Trial | Observations | Ordered Observations | Frequency |
|---|---|---|---|
| 45 | | | |
| 46 | | | |
| 47 | | | |
| 48 | | | |
| 49 | | | |
| 50 | | | |

Next group the data so that there are about 7 equal-interval classes and make a frequency table, showing the interval ranges and the frequencies in each class.

*Ranges*                                             *Frequencies*

Next graph the data from the frequency table. Make both a polygon and a histogram:

*Histogram*                                           *Polygon*

## THE NORMAL DISTRIBUTION

Next calculate the mean and standard deviation of the observed data.

$\bar{x} = $ _____

$s = $ _____

Draw vertical lines on the polygon to indicate the position of $\bar{x}$, of $\bar{x} + s$, and of $\bar{x} - s$, similar to that drawn for $\mu$ in Figure 26-4.

**Exercise 26-2:** Ask fifty people to tell you their heights to the nearest inch. Next put this raw data in ascending numerical order and fill in the following table.

| Number of Person | Observations | Ordered Observations | Frequency |
|---|---|---|---|
| 1 | | | |
| 2 | | | |
| 3 | | | |
| 4 | | | |
| 5 | | | |
| 6 | | | |
| 7 | | | |
| 8 | | | |
| 9 | | | |
| 10 | | | |
| 11 | | | |
| 12 | | | |
| 13 | | | |
| 14 | | | |
| 15 | | | |
| 16 | | | |
| 17 | | | |
| 18 | | | |
| 19 | | | |
| 20 | | | |
| 21 | | | |
| 22 | | | |
| 23 | | | |
| 24 | | | |
| 25 | | | |

## THE NORMAL DISTRIBUTION

| Number of Person | Observations | Ordered Observations | Frequency |
|---|---|---|---|
| 26 | | | |
| 27 | | | |
| 28 | | | |
| 29 | | | |
| 30 | | | |
| 31 | | | |
| 32 | | | |
| 33 | | | |
| 34 | | | |
| 35 | | | |
| 36 | | | |
| 37 | | | |
| 38 | | | |
| 39 | | | |
| 40 | | | |
| 41 | | | |
| 42 | | | |
| 43 | | | |
| 44 | | | |
| 45 | | | |
| 46 | | | |
| 47 | | | |
| 48 | | | |
| 49 | | | |
| 50 | | | |

## THE NORMAL DISTRIBUTION

Next group the data so that there are about 7 equal-interval ranges and list the frequencies in each class.

*Ranges*                          *Frequencies*

Next graph the data from the frequency table. Make both a polygon and a histogram.

*Histogram*                       *Polygon*

## THE NORMAL DISTRIBUTION

Next calculate the mean and standard deviation of the observed data.

$\bar{x} = $ _____

$s = $ _____

Draw vertical lines on the polygon to indicate the position of $\bar{x}$, of $\bar{x} + s$, and of $\bar{x} - s$, similar to that drawn for $\mu$ in Figure 26–4.

### TERMS TO REMEMBER

**Normal Distribution:** A commonly used distribution of observations whose graph is a smooth, symmetric, bell-shaped curve.

**Central Limit Theorem:** Given a population with mean $\mu$ and standard deviation $\sigma$, the distribution of the means of all possible samples of size $n$ that could be drawn from this population will, if $n$ is large enough, tend toward a normal distribution with mean $\mu$ and standard deviation $\dfrac{\sigma}{\sqrt{n}}$.

chapter 27

# CASE STUDIES

For the exercises of this chapter, use the methods presented in the preceding chapters to summarize the data, present the data in the most meaningful manner possible, and apply the appropriate statistical tests to complete the analysis.

***Exercise 27-1:*** A psychologist wishes to measure the degree of hostility among engineering students. Drawing a random sample of 125 engineering students, he presents to each student ten picture cards depicting people in ambiguous situations and asks him to create a story about the picture. He counts the number of aggression themes that appear in each student's ten stories and assigns this number as his hostility score. The following scores were observed:

1, 8, 7, 6, 3, 4, 4, 4, 4, 6, 7, 3, 3, 4, 4, 3, 1, 2, 8, 3, 4, 5,
4, 5, 8, 3, 4, 5, 4, 7, 4, 4, 5, 5, 5, 4, 6, 7, 6, 8, 6, 2, 6, 4,
7, 3, 4, 6, 5, 4, 3, 2, 1, 3, 5, 3, 6, 5, 6, 6, 4, 1, 2, 7, 5, 5,
5, 5, 4, 3, 2, 7, 5, 5, 4, 5, 7, 5, 4, 2, 5, 4, 4, 3, 2, 4, 3, 4,
3, 6, 4, 2, 2, 5, 6, 4, 7, 2, 4, 5, 5, 2, 2, 6, 4, 7, 5, 4, 6, 5,
4, 3, 1, 5, 6, 5, 4, 5, 2, 6, 8, 9, 5, 4, 6.

## CASE STUDIES

Use whatever methods you feel appropriate to analyze the data.

***Exercise 27-2:*** An automobile manufacturer has gathered data indicating the number of cars sold during the month of January that were equipped with air conditioning. The data have been gathered separately from each of 100 dealers. Use whatever methods you feel appropriate to analyze the data.

2, 4, 13, 6, 0, 7, 10, 8, 5, 7, 6, 4, 10, 5, 7, 6, 7, 5, 4, 5,
5, 6, 8, 10, 12, 1, 5, 4, 6, 3, 5, 6, 4, 2, 9, 7, 5, 5, 3, 1,
4, 7, 12, 2, 5, 6, 4, 5, 6, 3, 5, 5, 4, 1, 5, 8, 4, 6, 9, 2,
3, 5, 1, 7, 10, 8, 5, 4, 5, 6, 10, 2, 3, 5, 6, 7, 4, 1, 3, 10,
1, 5, 3, 6, 9, 0, 4, 7, 1, 5, 5, 6, 0, 2, 3, 4, 5, 6, 10, 5.

***Exercise 27-3:*** A chemical company has developed a new insecticide control for chinchbugs. They wish to compare its effectiveness with an existing product already being marketed. They have twenty containers, each with ten chinchbugs. Half of these containers are sprayed with the new product and half with the old product, and then the number of dead bugs per container is observed as follows:

| New product | 6 | 5 | 4 | 8 | 7 | 9 | 6 | 5 | 10 | 9 |
| Old product | 4 | 3 | 6 | 5 | 4 | 4 | 3 | 8 | 4 | 5 |

Analyze the data fully.

# CASE STUDIES

**Exercise 27-4:** A psychologist wants to measure how deeply juvenile delinquents feel a need for recognition. He presents to a randomly selected group of 50 juvenile delinquents a series of 10 pictures depicting people in ambiguous situations. Then he asks each to create a story about the situation. He counts the number of times the subjects relate stories with a desire or attempt to gain recognition as a theme. He then performs the same experiment with a group of 50 randomly selected nondelinquent teenagers. The data are summarized in the frequency table below. Use whatever methods you feel appropriate to analyze the data.

THEMES OF RECOGNITION-SEEKING

| No. of Recognition Themes | Delinquents | Non-Delinquents |
|---|---|---|
| 10 | 3 | 4 |
| 9 | 5 | 7 |
| 8 | 3 | 1 |
| 7 | 1 | 1 |
| 6 | 0 | 0 |
| 5 | 0 | 1 |
| 4 | 1 | 2 |
| 3 | 3 | 5 |
| 2 | 8 | 10 |
| 1 | 20 | 12 |
| 0 | 6 | 7 |
|  | 50 | 50 |

**Exercise 27–5:** In a genetics experiment a *w*hite-eyed, *l*ong-winged *m*ale fly (abbreviated WLM) is crossed with a *r*ed-eyed, *v*estigial-winged *f*emale fly (abbreviated RVF) to produce a generation of red-eyed, long-winged offspring (since red eyes and long wings are the dominant traits). These offspring are then crossed among themselves. It would be expected that their offspring would have the following proportions for each of the characteristics. The first letter of the abbreviation stands for eye color, the second for wing length, and the third for sex:

RLF, 6/16; RVF, 2/16; RLM, 3/16; RVM, 1/16; WLM, 3/16; WVM, 1/16.

In the 320 offspring produced, the following frequencies were observed for each characteristic: RLF, 115; RVF, 48; RLM, 54; RVM, 18; WLM, 60; and WVM, 25. Did the offspring follow the expected pattern?

**Exercise 27–6:** A sociologist has gathered data concerning 96 families, each of which has four children. One of the items of interest was the distribution of male and female children in each family. His data are summarized in the following table. Does there appear to be a 50% chance of getting either a boy or a girl? (Hint: The model would be a binomial distribution with $n = 4$ and $p = 0.5$.)

|  | 0 Boy<br>4 Girls | 1 Boy<br>3 Girls | 2 Boys<br>2 Girls | 3 Boys<br>1 Girl | 4 Boys<br>0 Girl |
|---|---|---|---|---|---|
| Frequency | 8 | 27 | 40 | 15 | 6 |

chapter 28

# A RESEARCH SURVEY

Forming groups of three or four members, each group should do the following:

(1) Decide on a questionnaire containing just two questions on any subject. Each group should use different questions.
(2) Design the questionnaire, phrasing the questions so that they are easily understood and are not in any manner ambiguous. Have one question requiring answers of the form yes–no–undecided, and one with answers of the form in favor of–opposed–no opinion.
(3) Ask these questions of all class members and record the data in the following table, so that for each person (identified by number only—no names) an answer to both questions is recorded.
(4) Before the next class, each group member should obtain answers from a sufficient number of persons so that each group study contains at least 40 observations with both questions answered by each subject.
(5) Analyze the data fully.
(6) As a final test, calculate a $\chi^2$ value indicating whether the answers to the two questions are related. To do this, delete the "undecided" and "no opinion" categories and set up a $2 \times 2$ table for the $\chi^2$ test with the yes-no versus the in favor of–opposed answers.

Use Table 28–1 to record the data.

**TABLE 28-1** DATA TABLE

| Respondee | Question 1 ||| Question 2 |||
|---|---|---|---|---|---|---|
| | YES | NO | UNDECIDED | IN FAVOR OF | OPPOSED | NO OPINION |
| 1 | | | | | | |
| 2 | | | | | | |
| 3 | | | | | | |
| 4 | | | | | | |
| 5 | | | | | | |
| 6 | | | | | | |
| 7 | | | | | | |
| 8 | | | | | | |
| 9 | | | | | | |
| 10 | | | | | | |
| 11 | | | | | | |
| 12 | | | | | | |
| 13 | | | | | | |
| 14 | | | | | | |
| 15 | | | | | | |
| 16 | | | | | | |
| 17 | | | | | | |
| 18 | | | | | | |
| 19 | | | | | | |
| 20 | | | | | | |
| 21 | | | | | | |
| 22 | | | | | | |
| 23 | | | | | | |
| 24 | | | | | | |
| 25 | | | | | | |
| 26 | | | | | | |
| 27 | | | | | | |
| 28 | | | | | | |
| 29 | | | | | | |
| 30 | | | | | | |
| 31 | | | | | | |
| 32 | | | | | | |
| 33 | | | | | | |
| 34 | | | | | | |
| 35 | | | | | | |
| 36 | | | | | | |
| 37 | | | | | | |
| 38 | | | | | | |
| 39 | | | | | | |
| 40 | | | | | | |

# APPENDICES

# APPENDICES

***TABLE I*** *A TABLE OF RANDOM UNITS.*

| Row/Col (1) | (2) | (3) | (4) | (5) | (6) | (7) | (8) | (9) | (10) |
|---|---|---|---|---|---|---|---|---|---|
| 1 | 56095 | 69245 | 29189 | 58995 | 71694 | 23299 | 40417 | 07329 | 34289 | 85551 |
| 2 | 27629 | 26621 | 65252 | 54825 | 32668 | 92174 | 23152 | 66208 | 77833 | 31665 |
| 3 | 35886 | 67191 | 13366 | 96945 | 51272 | 60657 | 00888 | 86856 | 44954 | 15221 |
| 4 | 98457 | 21408 | 15405 | 95676 | 05867 | 99943 | 21693 | 54112 | 29181 | 14102 |
| 5 | 80524 | 05988 | 28599 | 37781 | 13393 | 71766 | 77770 | 51794 | 97662 | 43199 |
| 6 | 67726 | 66447 | 73201 | 62046 | 55921 | 65946 | 57103 | 24196 | 67916 | 24769 |
| 7 | 22626 | 21775 | 99150 | 61576 | 97760 | 72264 | 83691 | 81921 | 57671 | 47603 |
| 8 | 11590 | 98397 | 78710 | 19915 | 41899 | 41616 | 18272 | 42248 | 04331 | 62411 |
| 9 | 68000 | 71771 | 90069 | 59137 | 48652 | 76845 | 48735 | 76576 | 56727 | 46687 |
| 10 | 84653 | 04633 | 57101 | 11561 | 22664 | 34381 | 18555 | 37554 | 86924 | 11777 |
| 11 | 28629 | 06740 | 76150 | 25174 | 34063 | 24063 | 08381 | 85890 | 62664 | 17965 |
| 12 | 29331 | 33882 | 27527 | 74398 | 55449 | 67167 | 99255 | 49869 | 41332 | 89830 |
| 13 | 74609 | 08110 | 52446 | 88430 | 78442 | 59881 | 38587 | 07200 | 83539 | 84344 |
| 14 | 85655 | 09542 | 21294 | 05688 | 05156 | 57634 | 17958 | 89285 | 24704 | 08219 |
| 15 | 27579 | 43340 | 39653 | 27354 | 68071 | 61822 | 20863 | 53469 | 55707 | 91447 |
| 16 | 22304 | 75029 | 92990 | 88816 | 13899 | 95588 | 72633 | 24662 | 45247 | 17658 |
| 17 | 28964 | 57666 | 11855 | 08430 | 96055 | 89097 | 52588 | 84799 | 68849 | 99077 |
| 18 | 03790 | 04229 | 70921 | 38310 | 81943 | 62685 | 75494 | 47394 | 22634 | 09569 |
| 19 | 51523 | 80028 | 16242 | 65809 | 52501 | 55680 | 69476 | 38247 | 50326 | 08041 |
| 20 | 92786 | 66873 | 21919 | 98074 | 45585 | 58832 | 17489 | 25493 | 61603 | 67193 |
| 21 | 51668 | 56797 | 68927 | 13652 | 64342 | 14850 | 19119 | 13895 | 68017 | 63560 |
| 22 | 91240 | 94972 | 50382 | 59112 | 32241 | 22269 | 62007 | 77588 | 62487 | 71721 |
| 23 | 85531 | 28951 | 12649 | 72272 | 40126 | 44776 | 00860 | 42236 | 37172 | 82183 |
| 24 | 17310 | 02397 | 58693 | 11552 | 81681 | 36781 | 50829 | 88019 | 36204 | 16364 |
| 25 | 98958 | 63695 | 54462 | 39529 | 80337 | 29569 | 16153 | 98901 | 16254 | 03733 |
| 26 | 34806 | 65100 | 64644 | 47318 | 91606 | 52122 | 99086 | 17538 | 96654 | 85210 |
| 27 | 78006 | 40624 | 23992 | 65646 | 52332 | 80715 | 89947 | 38817 | 37117 | 72414 |
| 28 | 25976 | 59669 | 26166 | 45186 | 34423 | 21705 | 14726 | 15520 | 19727 | 04032 |
| 29 | 92867 | 01286 | 30681 | 03863 | 25475 | 02284 | 67599 | 39253 | 47617 | 85886 |
| 30 | 91747 | 96502 | 04113 | 00240 | 96732 | 73085 | 41236 | 48136 | 57824 | 94929 |
| 31 | 71167 | 06642 | 32596 | 04627 | 41706 | 54520 | 47614 | 06291 | 37914 | 89485 |
| 32 | 53748 | 17374 | 44728 | 53417 | 07124 | 02532 | 15091 | 39933 | 21324 | 21670 |
| 33 | 41919 | 37415 | 61395 | 45715 | 33665 | 77855 | 19485 | 85081 | 39478 | 24643 |
| 34 | 38451 | 17526 | 11263 | 09197 | 62868 | 77465 | 59565 | 32467 | 99979 | 27002 |
| 35 | 89755 | 44966 | 25527 | 14059 | 88285 | 99317 | 23514 | 84840 | 92853 | 15455 |
| 36 | 11194 | 23577 | 96043 | 28471 | 77914 | 39515 | 36212 | 36199 | 97318 | 20831 |
| 37 | 73961 | 88783 | 93785 | 39157 | 42775 | 25784 | 60459 | 44823 | 36509 | 46262 |
| 38 | 17782 | 45050 | 61945 | 47760 | 20015 | 29391 | 81921 | 55352 | 68538 | 72959 |
| 39 | 30762 | 21991 | 25579 | 76661 | 61264 | 27674 | 77043 | 38112 | 47343 | 56537 |
| 40 | 35629 | 11014 | 23303 | 96793 | 78478 | 27039 | 04667 | 75337 | 30489 | 66695 |
| 41 | 77271 | 15332 | 75643 | 96354 | 03479 | 28834 | 93852 | 26440 | 45765 | 34213 |
| 42 | 43517 | 15651 | 65333 | 65340 | 35413 | 64752 | 36655 | 16166 | 81473 | 00285 |
| 43 | 50447 | 43179 | 13509 | 89675 | 44458 | 24349 | 97053 | 97720 | 17882 | 67550 |
| 44 | 14612 | 87769 | 60699 | 44399 | 11446 | 57091 | 02370 | 36136 | 75632 | 88254 |
| 45 | 92781 | 24214 | 84273 | 32051 | 12178 | 26252 | 86059 | 69540 | 88368 | 64831 |
| 46 | 43455 | 87358 | 55602 | 63412 | 31046 | 90211 | 55232 | 53785 | 86373 | 73579 |
| 47 | 74934 | 23492 | 96757 | 59228 | 59412 | 04714 | 78585 | 13196 | 60711 | 07277 |
| 48 | 12383 | 33078 | 20553 | 86557 | 13164 | 79380 | 33323 | 34578 | 57717 | 36343 |
| 49 | 51611 | 95281 | 40585 | 87070 | 58244 | 51850 | 76794 | 01150 | 77886 | 73741 |
| 50 | 12015 | 62488 | 97612 | 61909 | 30052 | 88326 | 46701 | 96626 | 44791 | 87846 |

EXAMPLE 1. We may use the table to simulate the tossing of a coin; say "even" is "heads" and "odd" is "tails". We can start anywhere in the table and read left, right, up, or down. Suppose we begin in row 6, column 3; then we can read (across to the right) 73201 62046 55921 65946. If we wish to simulate tossing a coin 20 times, we obtain TTHHT HHHHH TTTHT HTTHH, or 9 T and 11 H.

EXAMPLE 2. If we want to assign 5 rats to each of 3 experimental treatments, A, B, and C, we could number the rats "one" through "fifteen" and then obtain numbers from the table. If the table number is 1, 2, or 3, assign that rat to A; if 4, 5, or 6, assign it to B; if 7, 8, or 9, assign it to C; and if 0, skip that digit and go to the next. Beginning then in row 13, column 7, we read 38587 07200 83539 84344. Thus the rats would be assigned as follows:

| Rat Number | 1 2 3 4 5 | 6 7 | 8 9 10 11 | 12 13 14 15 |
|---|---|---|---|---|
| Assigned to treatment labeled | A C B C C | C A | C A B A | B A B B |

The vertical lines separate the assignments made from each group of five random digits. Notice that as soon as 5 rats were assigned to treatment "C", the numbers 7, 8, and 9 were then skipped when encountered.

*TABLE II CRITICAL VALUES TO TEST FOR RANDOMNESS IN SEQUENCES. NUMBERS OF RUNS GIVEN IN TABLE (OR VALUES MORE EXTREME) WOULD OCCUR IN A RANDOM SEQUENCE LESS THAN 10% OF THE TIME.*

| $N_2/N_1$ | 2 | 3 | 4 | 5 | 6 | 7 | 8 | 9 | 10 |
|---|---|---|---|---|---|---|---|---|---|
| 2 | | | | | | | (2,5) | (2,5) | (2,5) |
| 3 | | | | (2,7) | (2,7) | (2,7) | (2,7) | (2,7) | (3,7) |
| 4 | | | (2,7) | (2,8) | (3,8) | (3,8) | (3,9) | (3,9) | (3,9) |
| 5 | | (2,7) | (2,8) | (3,8) | (3,9) | (3,9) | (3,10) | (4,10) | (4,10) |
| 6 | | (2,7) | (3,8) | (3,9) | (3,10) | (4,10) | (4,11) | (4,11) | (5,11) |
| 7 | | (2,7) | (3,8) | (3,9) | (4,10) | (4,11) | (4,12) | (5,12) | (5,12) |
| 8 | (2,5) | (2,7) | (3,9) | (3,10) | (4,11) | (4,12) | (5,12) | (5,13) | (6,13) |
| 9 | (2,5) | (2,7) | (3,9) | (4,10) | (4,11) | (5,12) | (5,13) | (6,13) | (6,14) |
| 10 | (2,5) | (3,7) | (3,9) | (4,10) | (5,11) | (5,12) | (6,13) | (6,14) | (6,15) |

EXAMPLE. To test whether the following sequence is random:

<u>SS</u> <u>F</u> <u>SSS</u> <u>FF</u> <u>S</u> <u>F</u> <u>S</u> <u>FFFF</u>,

notice that the number of S's = $N_1$ = 7 and the number of F's = $N_2$ = 8, while the number of runs = 8. From the seventh column and eighth row of the table we read "(4,12)"; thus, since "8" lies between 4 and 12, we may conclude that the sample is random. If the number of runs observed had been 1, 2, 3, or 4, or if it had been 12, 13, 14, or 15, then

# APPENDICES

we would have concluded that the sequence was nonrandom, since a random sequence would have contained those particular numbers of runs less than ten per cent of the time.

**TABLE III** TABLE OF BINOMIAL PROBABILITIES.

$N$ = the number of binomial trials; $n$ = the number of successes in $N$ trials; and $p = P$(success) for each trial.

|   |   | p |   |   |   |   |
|---|---|---|---|---|---|---|
| N | n | .1 | .2 | .3 | .4 | .5 |
| 2 | 0 | .8100 | .6400 | .4900 | .3600 | .2500 |
|   | 1 | .1800 | .3200 | .4200 | .4800 | .5000 |
|   | 2 | .0100 | .0400 | .0900 | .1600 | .2500 |
| 3 | 0 | .7290 | .5120 | .3430 | .2160 | .1250 |
|   | 1 | .2430 | .3840 | .4410 | .4320 | .3750 |
|   | 2 | .0270 | .0960 | .1890 | .2880 | .3750 |
|   | 3 | .0010 | .0080 | .0270 | .0640 | .1250 |
| 4 | 0 | .6561 | .4096 | .2401 | .1296 | .0625 |
|   | 1 | .2916 | .4096 | .4116 | .3456 | .2500 |
|   | 2 | .0486 | .1536 | .2646 | .3456 | .3750 |
|   | 3 | .0036 | .0256 | .0756 | .1536 | .2500 |
|   | 4 | .0001 | .0016 | .0081 | .0256 | .0625 |
| 5 | 0 | .5905 | .3277 | .1681 | .0778 | .0312 |
|   | 1 | .3280 | .4096 | .3602 | .2592 | .1562 |
|   | 2 | .0729 | .2048 | .3087 | .3456 | .3125 |
|   | 3 | .0081 | .0512 | .1323 | .2304 | .3125 |
|   | 4 | .0004 | .0064 | .0284 | .0768 | .1562 |
|   | 5 | .0000 | .0003 | .0024 | .0102 | .0312 |
| 6 | 0 | .5314 | .2621 | .1176 | .0467 | .0156 |
|   | 1 | .3543 | .3932 | .3025 | .1866 | .0938 |
|   | 2 | .0984 | .2458 | .3241 | .3110 | .2344 |
|   | 3 | .0146 | .0819 | .1852 | .2765 | .3125 |
|   | 4 | .0012 | .0154 | .0595 | .1382 | .2344 |
|   | 5 | .0001 | .0015 | .0102 | .0369 | .0938 |
|   | 6 | .0000 | .0001 | .0007 | .0041 | .0156 |
| 7 | 0 | .4783 | .2097 | .0824 | .0280 | .0078 |
|   | 1 | .3720 | .3670 | .2471 | .1306 | .0547 |
|   | 2 | .1240 | .2753 | .3177 | .2613 | .1641 |
|   | 3 | .0230 | .1147 | .2269 | .2903 | .2734 |
|   | 4 | .0026 | .0287 | .0972 | .1935 | .2734 |
|   | 5 | .0002 | .0043 | .0250 | .0774 | .1641 |
|   | 6 | .0000 | .0004 | .0036 | .0172 | .0547 |
|   | 7 | .0000 | .0000 | .0002 | .0016 | .0078 |

**TABLE III**  TABLE OF BINOMIAL PROBABILITIES. (Continued)

| N | n | p=.1 | .2 | .3 | .4 | .5 |
|---|---|------|------|------|------|------|
| 8 | 0 | .4305 | .1678 | .0576 | .0168 | .0039 |
|   | 1 | .3826 | .3355 | .1977 | .0896 | .0312 |
|   | 2 | .1488 | .2936 | .2965 | .2090 | .1094 |
|   | 3 | .0331 | .1468 | .2541 | .2787 | .2188 |
|   | 4 | .0046 | .0459 | .1361 | .2322 | .2734 |
|   | 5 | .0004 | .0092 | .0467 | .1239 | .2188 |
|   | 6 | .0000 | .0011 | .0100 | .0413 | .1094 |
|   | 7 | .0000 | .0001 | .0012 | .0079 | .0312 |
|   | 8 | .0000 | .0000 | .0001 | .0007 | .0039 |
| 9 | 0 | .3874 | .1342 | .0404 | .0101 | .0020 |
|   | 1 | .3874 | .3020 | .1556 | .0605 | .0176 |
|   | 2 | .1722 | .3020 | .2668 | .1612 | .0703 |
|   | 3 | .0446 | .1762 | .2668 | .2508 | .1641 |
|   | 4 | .0074 | .0661 | .1715 | .2508 | .2461 |
|   | 5 | .0008 | .0165 | .0735 | .1672 | .2461 |
|   | 6 | .0001 | .0028 | .0210 | .0743 | .1641 |
|   | 7 | .0000 | .0003 | .0039 | .0212 | .0703 |
|   | 8 | .0000 | .0000 | .0004 | .0035 | .0176 |
|   | 9 | .0000 | .0000 | .0000 | .0003 | .0020 |
| 10 | 0 | .3487 | .1074 | .0282 | .0060 | .0010 |
|   | 1 | .3874 | .2684 | .1211 | .0403 | .0098 |
|   | 2 | .1937 | .3020 | .2335 | .1209 | .0439 |
|   | 3 | .0574 | .2013 | .2668 | .2150 | .1172 |
|   | 4 | .0112 | .0881 | .2001 | .2508 | .2051 |
|   | 5 | .0015 | .0264 | .1029 | .2007 | .2461 |
|   | 6 | .0001 | .0055 | .0368 | .1115 | .2051 |
|   | 7 | .0000 | .0008 | .0090 | .0425 | .1172 |
|   | 8 | .0000 | .0001 | .0014 | .0106 | .0439 |
|   | 9 | .0000 | .0000 | .0001 | .0016 | .0098 |
|   | 10 | .0000 | .0000 | .0000 | .0001 | .0010 |

EXAMPLE 1. What are the chances of getting at least 3 heads when a fair coin is tossed 5 times? Since it is given that the coin is "fair," we enter the table at $p = .5$. Since the coin was tossed 5 times, $N = 5$. "At least 3 heads" means either 3, or 4, or 5 heads could have turned up; so we add the probabilities for $n = 3$, $n = 4$, and $n = 5$ to obtain $.3125 + .1562 + .0312 = .4999$.

EXAMPLE 2. What are the chances of getting two defective TV sets in a lot of six TV sets if the manufacturer specifies a 1/10 probability of defectives? Here $N = 6$, $n = 2$, and $p = .1$; from the table, the chances can be read as .0984.

**TABLE IV** CRITICAL VALUES OF T IN THE WILCOXON MATCHED-PAIRS SIGNED RANKS TEST. VALUES OF T LESS THAN OR EQUAL TO THOSE GIVEN IN THE TABLE WOULD OCCUR WITH PROBABILITY LESS THAN THE LEVEL INDICATED IF THE TWO MATCHED GROUPS SHOWED NO TRUE DIFFERENCES.*

| | | Probability Level | | |
|---|---|---|---|---|
| n | .10 | .05 | .02 | .01 |
| 5 | 1 | — | — | — |
| 6 | 2 | 1 | — | — |
| 7 | 4 | 2 | 0 | — |
| 8 | 6 | 4 | 2 | 0 |
| 9 | 8 | 6 | 3 | 2 |
| 10 | 11 | 8 | 5 | 3 |
| 11 | 14 | 11 | 7 | 5 |
| 12 | 17 | 14 | 10 | 7 |
| 13 | 21 | 17 | 13 | 10 |
| 14 | 26 | 21 | 16 | 13 |
| 15 | 30 | 25 | 20 | 16 |
| 20 | 60 | 52 | 43 | 37 |
| 25 | 101 | 90 | 77 | 68 |
| 30 | 152 | 137 | 120 | 109 |
| 35 | 214 | 195 | 174 | 160 |
| 40 | 287 | 264 | 238 | 221 |
| 45 | 371 | 344 | 313 | 292 |
| 50 | 466 | 434 | 398 | 373 |

*Taken from "Some Rapid Approximate Statistical Procedures" by Frank Wilcoxon and Roberta A. Wilcox, published by Lederle Laboratories, Pearl River, New York, 1964. Printed by permission of the authors and Lederle Laboratories.

EXAMPLE. Assume the following observations for matched pairs with at least ordinal measurement scale:

| Group A | 23 | 36 | 15 | 42 | 15 | 29 | 27 |
|---|---|---|---|---|---|---|---|
| Group B | 24 | 38 | 15 | 41 | 19 | 31 | 28 |
| Differences | −1 | −2 | 0 | +1 | −4 | −2 | −1 |
| Ranks | 2 | 4.5 | — | 2 | 6 | 4.5 | 2 |
| Signed Ranks | −2 | −4.5 | | +2 | −6 | −4.5 | −2 |

T = 2

The critical value for $n = 6$ is 2 at the probability level .10. Thus we conclude that a value this small would occur by chance less than 10% of the time if there were no differences between Groups A and B.

# APPENDICES

**TABLE V** CRITICAL VALUES ASSOCIATED WITH THE MANN-WHITNEY U TEST. VALUES OF U LESS THAN OR EQUAL TO THOSE GIVEN IN THE TABLE WOULD OCCUR ONLY 10% OF THE TIME OR LESS IF THERE WERE NO TRUE DIFFERENCE IN THE TREATMENT EFFECTS OF THE TWO INDEPENDENT GROUPS OBSERVED.

| $n_2$ = Sample Size in Sample 2 | 3 | 4 | 5 | 6 | 7 | 8 | 9 | 10 | 11 | 12 | 13 | 14 | 15 | 16 | 17 | 18 | 19 | 20 |
|---|---|---|---|---|---|---|---|---|---|---|---|---|---|---|---|---|---|---|
| 3 | 0 | 0 | 1 | 2 | 2 | 3 | 3 | 4 | 5 | 5 | 6 | 7 | 7 | 8 | 9 | 9 | 10 | 11 |
| 4 | 0 | 1 | 2 | 3 | 4 | 5 | 6 | 7 | 8 | 9 | 10 | 11 | 12 | 14 | 15 | 16 | 17 | 18 |
| 5 | 1 | 2 | 4 | 5 | 6 | 8 | 9 | 11 | 12 | 13 | 15 | 16 | 18 | 19 | 20 | 22 | 23 | 25 |
| 6 | 2 | 3 | 5 | 7 | 8 | 10 | 12 | 14 | 16 | 17 | 19 | 21 | 23 | 25 | 26 | 28 | 30 | 32 |
| 7 | 2 | 4 | 6 | 8 | 11 | 13 | 15 | 17 | 19 | 21 | 24 | 26 | 28 | 30 | 33 | 35 | 37 | 39 |
| 8 | 3 | 5 | 8 | 10 | 13 | 15 | 18 | 20 | 23 | 26 | 28 | 31 | 33 | 36 | 39 | 41 | 44 | 47 |
| 9 | 3 | 6 | 9 | 12 | 15 | 18 | 21 | 24 | 27 | 30 | 33 | 36 | 39 | 42 | 45 | 48 | 51 | 54 |
| 10 | 4 | 7 | 11 | 14 | 17 | 20 | 24 | 27 | 31 | 34 | 37 | 41 | 44 | 48 | 51 | 55 | 58 | 62 |
| 11 | 5 | 8 | 12 | 16 | 19 | 23 | 27 | 31 | 34 | 38 | 42 | 46 | 50 | 54 | 57 | 61 | 65 | 69 |
| 12 | 5 | 9 | 13 | 17 | 21 | 26 | 30 | 34 | 38 | 42 | 47 | 51 | 55 | 60 | 64 | 68 | 72 | 77 |
| 13 | 6 | 10 | 15 | 19 | 24 | 28 | 33 | 37 | 42 | 47 | 51 | 56 | 61 | 65 | 70 | 75 | 80 | 84 |
| 14 | 7 | 11 | 16 | 21 | 26 | 31 | 36 | 41 | 46 | 51 | 56 | 61 | 66 | 71 | 77 | 82 | 87 | 92 |
| 15 | 7 | 12 | 18 | 23 | 28 | 33 | 39 | 44 | 50 | 55 | 61 | 66 | 72 | 77 | 83 | 88 | 94 | 100 |
| 16 | 8 | 14 | 19 | 25 | 30 | 36 | 42 | 48 | 54 | 60 | 65 | 71 | 77 | 83 | 89 | 95 | 101 | 107 |
| 17 | 9 | 15 | 20 | 26 | 33 | 39 | 45 | 51 | 57 | 64 | 70 | 77 | 83 | 89 | 96 | 102 | 109 | 115 |
| 18 | 9 | 16 | 22 | 28 | 35 | 41 | 48 | 55 | 61 | 68 | 75 | 82 | 88 | 95 | 102 | 109 | 116 | 123 |
| 19 | 10 | 17 | 23 | 30 | 37 | 44 | 51 | 58 | 65 | 72 | 80 | 87 | 94 | 101 | 109 | 116 | 123 | 130 |
| 20 | 11 | 18 | 25 | 32 | 39 | 47 | 54 | 62 | 69 | 77 | 84 | 92 | 100 | 107 | 115 | 123 | 130 | 138 |

Probability Level = .10
$n_1$ = Sample Size in Sample 1

EXAMPLE. The following data are for independent groups having at least ordinal measurement:

| Group 1 | 13 | 17 | 22 | 14 | 20 |
| Group 2 | 21 | 18 | 23 | 20 |

1. Listed measures (Group 1 underlined): <u>12</u>, <u>13</u>, <u>14</u>, <u>17</u>, 18, <u>20</u>, 20, 21, 23
2. $n_1 = 5$ and $n_2 = 4$
3. Ranks <u>1</u>, <u>2</u>, <u>3</u>, <u>4</u>, 5, <u>6.5</u>, 6.5, 8, 9.
4. $R_1 = 16.5$ and $R_2 = 28.5$
5. $U_1 = 20 + \frac{30}{2} - 16.5 = 18.5$ and $U_2 = 20 + \frac{20}{2} - 28.5 = 1.5$
   Check: $18.5 = 20 - 1.5$
6. $U = 1.5$
7. Critical value for $n_1 = 5$ and $n_2 = 4$ is 2. Conclusion: the two groups differ, since a value this small would occur by chance less than ten per cent of the time if in fact there were no true difference between the groups.

# APPENDICES

**TABLE VI** TABLE OF SQUARES AND SQUARE ROOTS.

| $N$ | $N^2$ | $\sqrt{N}$ | $\sqrt{10N}$ | $N$ | $N^2$ | $\sqrt{N}$ | $\sqrt{10N}$ |
|---|---|---|---|---|---|---|---|
| 1.00 | 1.00 | 1.00 | 3.16 | 1.45 | 2.10 | 1.20 | 3.81 |
| 1.01 | 1.02 | 1.00 | 3.18 | 1.46 | 2.13 | 1.21 | 3.82 |
| 1.02 | 1.04 | 1.01 | 3.19 | 1.47 | 2.16 | 1.21 | 3.83 |
| 1.03 | 1.06 | 1.01 | 3.21 | 1.48 | 2.19 | 1.22 | 3.85 |
| 1.04 | 1.08 | 1.02 | 3.22 | 1.49 | 2.22 | 1.22 | 3.86 |
| 1.05 | 1.10 | 1.02 | 3.24 | 1.50 | 2.25 | 1.22 | 3.87 |
| 1.06 | 1.12 | 1.03 | 3.26 | 1.51 | 2.28 | 1.23 | 3.89 |
| 1.07 | 1.14 | 1.03 | 3.27 | 1.52 | 2.31 | 1.23 | 3.90 |
| 1.08 | 1.17 | 1.04 | 3.29 | 1.53 | 2.34 | 1.24 | 3.91 |
| 1.09 | 1.19 | 1.04 | 3.30 | 1.54 | 2.37 | 1.24 | 3.92 |
| 1.10 | 1.21 | 1.05 | 3.32 | 1.55 | 2.40 | 1.24 | 3.94 |
| 1.11 | 1.23 | 1.05 | 3.33 | 1.56 | 2.43 | 1.25 | 3.95 |
| 1.12 | 1.25 | 1.06 | 3.35 | 1.57 | 2.46 | 1.25 | 3.96 |
| 1.13 | 1.28 | 1.06 | 3.36 | 1.58 | 2.50 | 1.26 | 3.97 |
| 1.14 | 1.30 | 1.07 | 3.38 | 1.59 | 2.53 | 1.26 | 3.99 |
| 1.15 | 1.32 | 1.07 | 3.39 | 1.60 | 2.56 | 1.26 | 4.00 |
| 1.16 | 1.35 | 1.08 | 3.41 | 1.61 | 2.59 | 1.27 | 4.01 |
| 1.17 | 1.37 | 1.08 | 3.42 | 1.62 | 2.62 | 1.27 | 4.02 |
| 1.18 | 1.39 | 1.09 | 3.44 | 1.63 | 2.66 | 1.28 | 4.04 |
| 1.19 | 1.42 | 1.09 | 3.45 | 1.64 | 2.69 | 1.28 | 4.05 |
| 1.20 | 1.44 | 1.10 | 3.46 | 1.65 | 2.72 | 1.28 | 4.06 |
| 1.21 | 1.46 | 1.10 | 3.48 | 1.66 | 2.76 | 1.29 | 4.07 |
| 1.22 | 1.49 | 1.10 | 3.49 | 1.67 | 2.79 | 1.29 | 4.09 |
| 1.23 | 1.51 | 1.11 | 3.51 | 1.68 | 2.82 | 1.30 | 4.10 |
| 1.24 | 1.54 | 1.11 | 3.52 | 1.69 | 2.86 | 1.30 | 4.11 |
| 1.25 | 1.56 | 1.12 | 3.54 | 1.70 | 2.89 | 1.30 | 4.12 |
| 1.26 | 1.59 | 1.12 | 3.55 | 1.71 | 2.92 | 1.31 | 4.14 |
| 1.27 | 1.61 | 1.13 | 3.56 | 1.72 | 2.96 | 1.31 | 4.15 |
| 1.28 | 1.64 | 1.13 | 3.58 | 1.73 | 2.99 | 1.32 | 4.16 |
| 1.29 | 1.66 | 1.14 | 3.59 | 1.74 | 3.03 | 1.32 | 4.17 |
| 1.30 | 1.69 | 1.14 | 3.61 | 1.75 | 3.06 | 1.32 | 4.18 |
| 1.31 | 1.72 | 1.14 | 3.62 | 1.76 | 3.10 | 1.33 | 4.20 |
| 1.32 | 1.74 | 1.15 | 3.63 | 1.77 | 3.13 | 1.33 | 4.21 |
| 1.33 | 1.77 | 1.15 | 3.65 | 1.78 | 3.17 | 1.33 | 4.22 |
| 1.34 | 1.80 | 1.16 | 3.66 | 1.79 | 3.20 | 1.34 | 4.23 |
| 1.35 | 1.82 | 1.16 | 3.67 | 1.80 | 3.24 | 1.34 | 4.24 |
| 1.36 | 1.85 | 1.17 | 3.69 | 1.81 | 3.28 | 1.35 | 4.25 |
| 1.37 | 1.88 | 1.17 | 3.70 | 1.82 | 3.31 | 1.35 | 4.27 |
| 1.38 | 1.90 | 1.17 | 3.71 | 1.83 | 3.35 | 1.35 | 4.28 |
| 1.39 | 1.93 | 1.18 | 3.73 | 1.84 | 3.39 | 1.36 | 4.29 |
| 1.40 | 1.96 | 1.18 | 3.74 | 1.85 | 3.42 | 1.36 | 4.30 |
| 1.41 | 1.99 | 1.19 | 3.75 | 1.86 | 3.46 | 1.36 | 4.31 |
| 1.42 | 2.02 | 1.19 | 3.77 | 1.87 | 3.50 | 1.37 | 4.32 |
| 1.43 | 2.04 | 1.20 | 3.78 | 1.88 | 3.53 | 1.37 | 4.34 |
| 1.44 | 2.07 | 1.20 | 3.79 | 1.89 | 3.57 | 1.37 | 4.35 |

**TABLE VI**  TABLE OF SQUARES AND SQUARE ROOTS. (Continued)

| N | $N^2$ | $\sqrt{N}$ | $\sqrt{10N}$ | N | $N^2$ | $\sqrt{N}$ | $\sqrt{10N}$ |
|---|---|---|---|---|---|---|---|
| 1.90 | 3.61 | 1.38 | 4.36 | 2.40 | 5.76 | 1.55 | 4.90 |
| 1.91 | 3.65 | 1.38 | 4.37 | 2.41 | 5.81 | 1.55 | 4.91 |
| 1.92 | 3.69 | 1.39 | 4.38 | 2.42 | 5.86 | 1.56 | 4.92 |
| 1.93 | 3.72 | 1.39 | 4.39 | 2.43 | 5.90 | 1.56 | 4.93 |
| 1.94 | 3.76 | 1.39 | 4.40 | 2.44 | 5.95 | 1.56 | 4.94 |
| 1.95 | 3.80 | 1.40 | 4.42 | 2.45 | 6.00 | 1.57 | 4.95 |
| 1.96 | 3.84 | 1.40 | 4.43 | 2.46 | 6.05 | 1.57 | 4.96 |
| 1.97 | 3.88 | 1.40 | 4.44 | 2.47 | 6.10 | 1.57 | 4.97 |
| 1.98 | 3.92 | 1.41 | 4.45 | 2.48 | 6.15 | 1.57 | 4.98 |
| 1.99 | 3.96 | 1.41 | 4.46 | 2.49 | 6.20 | 1.58 | 4.99 |
| 2.00 | 4.00 | 1.41 | 4.47 | 2.50 | 6.25 | 1.58 | 5.00 |
| 2.01 | 4.04 | 1.42 | 4.48 | 2.51 | 6.30 | 1.58 | 5.01 |
| 2.02 | 4.08 | 1.42 | 4.49 | 2.52 | 6.35 | 1.59 | 5.02 |
| 2.03 | 4.12 | 1.42 | 4.51 | 2.53 | 6.40 | 1.59 | 5.03 |
| 2.04 | 4.16 | 1.43 | 4.52 | 2.54 | 6.45 | 1.59 | 5.04 |
| 2.05 | 4.20 | 1.43 | 4.53 | 2.55 | 6.50 | 1.60 | 5.05 |
| 2.06 | 4.24 | 1.44 | 4.54 | 2.56 | 6.55 | 1.60 | 5.06 |
| 2.07 | 4.28 | 1.44 | 4.55 | 2.57 | 6.60 | 1.60 | 5.07 |
| 2.08 | 4.33 | 1.44 | 4.56 | 2.58 | 6.66 | 1.61 | 5.08 |
| 2.09 | 4.37 | 1.45 | 4.57 | 2.59 | 6.71 | 1.61 | 5.09 |
| 2.10 | 4.41 | 1.45 | 4.58 | 2.60 | 6.76 | 1.61 | 5.10 |
| 2.11 | 4.45 | 1.45 | 4.59 | 2.61 | 6.81 | 1.62 | 5.11 |
| 2.12 | 4.49 | 1.46 | 4.60 | 2.62 | 6.86 | 1.62 | 5.12 |
| 2.13 | 4.54 | 1.46 | 4.62 | 2.63 | 6.92 | 1.62 | 5.13 |
| 2.14 | 4.58 | 1.46 | 4.63 | 2.64 | 6.97 | 1.62 | 5.14 |
| 2.15 | 4.62 | 1.47 | 4.64 | 2.65 | 7.02 | 1.63 | 5.15 |
| 2.16 | 4.67 | 1.47 | 4.65 | 2.66 | 7.08 | 1.63 | 5.16 |
| 2.17 | 4.71 | 1.47 | 4.66 | 2.67 | 7.13 | 1.63 | 5.17 |
| 2.18 | 4.75 | 1.48 | 4.67 | 2.68 | 7.18 | 1.64 | 5.18 |
| 2.19 | 4.80 | 1.48 | 4.68 | 2.69 | 7.24 | 1.64 | 5.19 |
| 2.20 | 4.84 | 1.48 | 4.69 | 2.70 | 7.29 | 1.64 | 5.20 |
| 2.21 | 4.88 | 1.49 | 4.70 | 2.71 | 7.34 | 1.65 | 5.21 |
| 2.22 | 4.93 | 1.49 | 4.71 | 2.72 | 7.40 | 1.65 | 5.22 |
| 2.23 | 4.97 | 1.49 | 4.72 | 2.73 | 7.45 | 1.65 | 5.22 |
| 2.24 | 5.02 | 1.50 | 4.73 | 2.74 | 7.51 | 1.66 | 5.23 |
| 2.25 | 5.06 | 1.50 | 4.74 | 2.75 | 7.56 | 1.66 | 5.24 |
| 2.26 | 5.11 | 1.50 | 4.75 | 2.76 | 7.62 | 1.66 | 5.25 |
| 2.27 | 5.15 | 1.51 | 4.76 | 2.77 | 7.67 | 1.66 | 5.26 |
| 2.28 | 5.20 | 1.51 | 4.77 | 2.78 | 7.73 | 1.67 | 5.27 |
| 2.29 | 5.24 | 1.51 | 4.79 | 2.79 | 7.78 | 1.67 | 5.28 |
| 2.30 | 5.29 | 1.52 | 4.80 | 2.80 | 7.84 | 1.67 | 5.29 |
| 2.31 | 5.34 | 1.52 | 4.81 | 2.81 | 7.90 | 1.68 | 5.30 |
| 2.32 | 5.38 | 1.52 | 4.82 | 2.82 | 7.95 | 1.68 | 5.31 |
| 2.33 | 5.43 | 1.53 | 4.83 | 2.83 | 8.01 | 1.68 | 5.32 |
| 2.34 | 5.48 | 1.53 | 4.84 | 2.84 | 8.07 | 1.69 | 5.33 |
| 2.35 | 5.52 | 1.53 | 4.85 | 2.85 | 8.12 | 1.69 | 5.34 |
| 2.36 | 5.57 | 1.54 | 4.86 | 2.86 | 8.18 | 1.69 | 5.35 |
| 2.37 | 5.62 | 1.54 | 4.87 | 2.87 | 8.24 | 1.69 | 5.36 |
| 2.38 | 5.66 | 1.54 | 4.88 | 2.88 | 8.29 | 1.70 | 5.37 |
| 2.39 | 5.71 | 1.55 | 4.89 | 2.89 | 8.35 | 1.70 | 5.38 |

**TABLE VI**  TABLE OF SQUARES AND SQUARE ROOTS. (Continued)

| N | N² | √N | √10N | N | N² | √N | √10N |
|---|---|---|---|---|---|---|---|
| 2.90 | 8.41 | 1.70 | 5.39 | 3.40 | 11.56 | 1.84 | 5.83 |
| 2.91 | 8.47 | 1.71 | 5.39 | 3.41 | 11.63 | 1.85 | 5.84 |
| 2.92 | 8.53 | 1.71 | 5.40 | 3.42 | 11.70 | 1.85 | 5.85 |
| 2.93 | 8.58 | 1.71 | 5.41 | 3.43 | 11.76 | 1.85 | 5.86 |
| 2.94 | 8.64 | 1.71 | 5.42 | 3.44 | 11.83 | 1.85 | 5.87 |
| 2.95 | 8.70 | 1.72 | 5.43 | 3.45 | 11.90 | 1.86 | 5.87 |
| 2.96 | 8.76 | 1.72 | 5.44 | 3.46 | 11.97 | 1.86 | 5.88 |
| 2.97 | 8.82 | 1.72 | 5.45 | 3.47 | 12.04 | 1.86 | 5.89 |
| 2.98 | 8.88 | 1.73 | 5.46 | 3.48 | 12.11 | 1.87 | 5.90 |
| 2.99 | 8.94 | 1.73 | 5.47 | 3.49 | 12.18 | 1.87 | 5.91 |
| 3.00 | 9.00 | 1.73 | 5.48 | 3.50 | 12.25 | 1.87 | 5.92 |
| 3.01 | 9.06 | 1.73 | 5.49 | 3.51 | 12.32 | 1.87 | 5.92 |
| 3.02 | 9.12 | 1.74 | 5.50 | 3.52 | 12.39 | 1.88 | 5.93 |
| 3.03 | 9.18 | 1.74 | 5.50 | 3.53 | 12.46 | 1.88 | 5.94 |
| 3.04 | 9.24 | 1.74 | 5.51 | 3.54 | 12.53 | 1.88 | 5.95 |
| 3.05 | 9.30 | 1.75 | 5.52 | 3.55 | 12.60 | 1.88 | 5.96 |
| 3.06 | 9.36 | 1.75 | 5.53 | 3.56 | 12.67 | 1.89 | 5.97 |
| 3.07 | 9.42 | 1.75 | 5.54 | 3.57 | 12.74 | 1.89 | 5.97 |
| 3.08 | 9.49 | 1.75 | 5.55 | 3.58 | 12.82 | 1.89 | 5.98 |
| 3.09 | 9.55 | 1.76 | 5.56 | 3.59 | 12.89 | 1.89 | 5.99 |
| 3.10 | 9.61 | 1.76 | 5.57 | 3.60 | 12.96 | 1.90 | 6.00 |
| 3.11 | 9.67 | 1.76 | 5.58 | 3.61 | 13.03 | 1.90 | 6.01 |
| 3.12 | 9.73 | 1.77 | 5.59 | 3.62 | 13.10 | 1.90 | 6.02 |
| 3.13 | 9.80 | 1.77 | 5.59 | 3.63 | 13.18 | 1.91 | 6.02 |
| 3.14 | 9.86 | 1.77 | 5.60 | 3.64 | 13.25 | 1.91 | 6.03 |
| 3.15 | 9.92 | 1.77 | 5.61 | 3.65 | 13.32 | 1.91 | 6.04 |
| 3.16 | 9.99 | 1.78 | 5.62 | 3.66 | 13.40 | 1.91 | 6.05 |
| 3.17 | 10.05 | 1.78 | 5.63 | 3.67 | 13.47 | 1.92 | 6.06 |
| 3.18 | 10.11 | 1.78 | 5.64 | 3.68 | 13.54 | 1.92 | 6.07 |
| 3.19 | 10.18 | 1.79 | 5.65 | 3.69 | 13.62 | 1.92 | 6.07 |
| 3.20 | 10.24 | 1.79 | 5.66 | 3.70 | 13.69 | 1.92 | 6.08 |
| 3.21 | 10.30 | 1.79 | 5.67 | 3.71 | 13.76 | 1.93 | 6.09 |
| 3.22 | 10.37 | 1.79 | 5.67 | 3.72 | 13.84 | 1.93 | 6.10 |
| 3.23 | 10.43 | 1.80 | 5.68 | 3.73 | 13.91 | 1.93 | 6.11 |
| 3.24 | 10.50 | 1.80 | 5.69 | 3.74 | 13.99 | 1.93 | 6.12 |
| 3.25 | 10.56 | 1.80 | 5.70 | 3.75 | 14.06 | 1.94 | 6.12 |
| 3.26 | 10.63 | 1.81 | 5.71 | 3.76 | 14.14 | 1.94 | 6.13 |
| 3.27 | 10.69 | 1.81 | 5.72 | 3.77 | 14.21 | 1.94 | 6.14 |
| 3.28 | 10.76 | 1.81 | 5.73 | 3.78 | 14.29 | 1.94 | 6.15 |
| 3.29 | 10.82 | 1.81 | 5.74 | 3.79 | 14.36 | 1.95 | 6.16 |
| 3.30 | 10.89 | 1.82 | 5.74 | 3.80 | 14.44 | 1.95 | 6.16 |
| 3.31 | 10.96 | 1.82 | 5.75 | 3.81 | 14.52 | 1.95 | 6.17 |
| 3.32 | 11.02 | 1.82 | 5.76 | 3.82 | 14.59 | 1.95 | 6.18 |
| 3.33 | 11.09 | 1.82 | 5.77 | 3.83 | 14.67 | 1.96 | 6.19 |
| 3.34 | 11.16 | 1.83 | 5.78 | 3.84 | 14.75 | 1.96 | 6.20 |
| 3.35 | 11.22 | 1.83 | 5.79 | 3.85 | 14.82 | 1.96 | 6.20 |
| 3.36 | 11.29 | 1.83 | 5.80 | 3.86 | 14.90 | 1.96 | 6.21 |
| 3.37 | 11.36 | 1.84 | 5.81 | 3.87 | 14.98 | 1.97 | 6.22 |
| 3.38 | 11.42 | 1.84 | 5.81 | 3.88 | 15.05 | 1.97 | 6.23 |
| 3.39 | 11.49 | 1.84 | 5.82 | 3.89 | 15.13 | 1.97 | 6.24 |

**TABLE VI** TABLE OF SQUARES AND SQUARE ROOTS. (Continued)

| N | $N^2$ | $\sqrt{N}$ | $\sqrt{10N}$ | N | $N^2$ | $\sqrt{N}$ | $\sqrt{10N}$ |
|---|---|---|---|---|---|---|---|
| 3.90 | 15.21 | 1.97 | 6.24 | 4.40 | 19.36 | 2.10 | 6.63 |
| 3.91 | 15.29 | 1.98 | 6.25 | 4.41 | 19.45 | 2.10 | 6.64 |
| 3.92 | 15.37 | 1.98 | 6.26 | 4.42 | 19.54 | 2.10 | 6.65 |
| 3.93 | 15.44 | 1.98 | 6.27 | 4.43 | 19.62 | 2.10 | 6.66 |
| 3.94 | 15.52 | 1.98 | 6.28 | 4.44 | 19.71 | 2.11 | 6.66 |
| 3.95 | 15.60 | 1.99 | 6.28 | 4.45 | 19.80 | 2.11 | 6.67 |
| 3.96 | 15.68 | 1.99 | 6.29 | 4.46 | 19.89 | 2.11 | 6.68 |
| 3.97 | 15.76 | 1.99 | 6.30 | 4.47 | 19.98 | 2.11 | 6.69 |
| 3.98 | 15.84 | 1.99 | 6.31 | 4.48 | 20.07 | 2.12 | 6.69 |
| 3.99 | 15.92 | 2.00 | 6.32 | 4.49 | 20.16 | 2.12 | 6.70 |
| 4.00 | 16.00 | 2.00 | 6.32 | 4.50 | 20.25 | 2.12 | 6.71 |
| 4.01 | 16.08 | 2.00 | 6.33 | 4.51 | 20.34 | 2.12 | 6.72 |
| 4.02 | 16.16 | 2.00 | 6.34 | 4.52 | 20.43 | 2.13 | 6.72 |
| 4.03 | 16.24 | 2.01 | 6.35 | 4.53 | 20.52 | 2.13 | 6.73 |
| 4.04 | 16.32 | 2.01 | 6.36 | 4.54 | 20.61 | 2.13 | 6.74 |
| 4.05 | 16.40 | 2.01 | 6.36 | 4.55 | 20.70 | 2.13 | 6.75 |
| 4.06 | 16.48 | 2.01 | 6.37 | 4.56 | 20.79 | 2.14 | 6.75 |
| 4.07 | 16.56 | 2.02 | 6.38 | 4.57 | 20.88 | 2.14 | 6.76 |
| 4.08 | 16.65 | 2.02 | 6.39 | 4.58 | 20.98 | 2.14 | 6.77 |
| 4.09 | 16.73 | 2.02 | 6.40 | 4.59 | 21.07 | 2.14 | 6.77 |
| 4.10 | 16.81 | 2.02 | 6.40 | 4.60 | 21.16 | 2.14 | 6.78 |
| 4.11 | 16.89 | 2.03 | 6.41 | 4.61 | 21.25 | 2.15 | 6.79 |
| 4.12 | 16.97 | 2.03 | 6.42 | 4.62 | 21.34 | 2.15 | 6.80 |
| 4.13 | 17.06 | 2.03 | 6.43 | 4.63 | 21.44 | 2.15 | 6.80 |
| 4.14 | 17.14 | 2.03 | 6.43 | 4.64 | 21.53 | 2.15 | 6.81 |
| 4.15 | 17.22 | 2.04 | 6.44 | 4.65 | 21.62 | 2.16 | 6.82 |
| 4.16 | 17.31 | 2.04 | 6.45 | 4.66 | 21.72 | 2.16 | 6.83 |
| 4.17 | 17.39 | 2.04 | 6.46 | 4.67 | 21.81 | 2.16 | 6.83 |
| 4.18 | 17.47 | 2.04 | 6.47 | 4.68 | 21.90 | 2.16 | 6.84 |
| 4.19 | 17.56 | 2.05 | 6.47 | 4.69 | 22.00 | 2.17 | 6.85 |
| 4.20 | 17.64 | 2.05 | 6.48 | 4.70 | 22.09 | 2.17 | 6.86 |
| 4.21 | 17.72 | 2.05 | 6.49 | 4.71 | 22.18 | 2.17 | 6.86 |
| 4.22 | 17.81 | 2.05 | 6.50 | 4.72 | 22.28 | 2.17 | 6.87 |
| 4.23 | 17.89 | 2.06 | 6.50 | 4.73 | 22.37 | 2.17 | 6.88 |
| 4.24 | 17.98 | 2.06 | 6.51 | 4.74 | 22.47 | 2.18 | 6.88 |
| 4.25 | 18.06 | 2.06 | 6.52 | 4.75 | 22.56 | 2.18 | 6.89 |
| 4.26 | 18.15 | 2.06 | 6.53 | 4.76 | 22.66 | 2.18 | 6.90 |
| 4.27 | 18.23 | 2.07 | 6.53 | 4.77 | 22.75 | 2.18 | 6.91 |
| 4.28 | 18.32 | 2.07 | 6.54 | 4.78 | 22.85 | 2.19 | 6.91 |
| 4.29 | 18.40 | 2.07 | 6.55 | 4.79 | 22.94 | 2.19 | 6.92 |
| 4.30 | 18.49 | 2.07 | 6.56 | 4.80 | 23.04 | 2.19 | 6.93 |
| 4.31 | 18.58 | 2.08 | 6.57 | 4.81 | 23.14 | 2.19 | 6.94 |
| 4.32 | 18.66 | 2.08 | 6.57 | 4.82 | 23.23 | 2.20 | 6.94 |
| 4.33 | 18.75 | 2.08 | 6.58 | 4.83 | 23.33 | 2.20 | 6.95 |
| 4.34 | 18.84 | 2.08 | 6.59 | 4.84 | 23.43 | 2.20 | 6.96 |
| 4.35 | 18.92 | 2.09 | 6.60 | 4.85 | 23.52 | 2.20 | 6.96 |
| 4.36 | 19.01 | 2.09 | 6.60 | 4.86 | 23.62 | 2.20 | 6.97 |
| 4.37 | 19.10 | 2.09 | 6.61 | 4.87 | 23.72 | 2.21 | 6.98 |
| 4.38 | 19.18 | 2.09 | 6.62 | 4.88 | 23.81 | 2.21 | 6.99 |
| 4.39 | 19.27 | 2.10 | 6.63 | 4.89 | 23.91 | 2.21 | 6.99 |

## APPENDICES

**TABLE VI** TABLE OF SQUARES AND SQUARE ROOTS. (Continued)

| N | N² | √N | √10N | N | N² | √N | √10N |
|---|---|---|---|---|---|---|---|
| 4.90 | 24.01 | 2.21 | 7.00 | 5.40 | 29.16 | 2.32 | 7.35 |
| 4.91 | 24.11 | 2.22 | 7.01 | 5.41 | 29.27 | 2.33 | 7.36 |
| 4.92 | 24.21 | 2.22 | 7.01 | 5.42 | 29.38 | 2.33 | 7.36 |
| 4.93 | 24.30 | 2.22 | 7.02 | 5.43 | 29.48 | 2.33 | 7.37 |
| 4.94 | 24.40 | 2.22 | 7.03 | 5.44 | 29.59 | 2.33 | 7.38 |
| 4.95 | 24.50 | 2.22 | 7.04 | 5.45 | 29.70 | 2.33 | 7.38 |
| 4.96 | 24.60 | 2.23 | 7.04 | 5.46 | 29.81 | 2.34 | 7.39 |
| 4.97 | 24.70 | 2.23 | 7.05 | 5.47 | 29.92 | 2.34 | 7.40 |
| 4.98 | 24.80 | 2.23 | 7.06 | 5.48 | 30.03 | 2.34 | 7.40 |
| 4.99 | 24.90 | 2.23 | 7.06 | 5.49 | 30.14 | 2.34 | 7.41 |
| 5.00 | 25.00 | 2.24 | 7.07 | 5.50 | 30.25 | 2.35 | 7.42 |
| 5.01 | 25.10 | 2.24 | 7.08 | 5.51 | 30.36 | 2.35 | 7.42 |
| 5.02 | 25.20 | 2.24 | 7.09 | 5.52 | 30.47 | 2.35 | 7.43 |
| 5.03 | 25.30 | 2.24 | 7.09 | 5.53 | 30.58 | 2.35 | 7.44 |
| 5.04 | 25.40 | 2.24 | 7.10 | 5.54 | 30.69 | 2.35 | 7.44 |
| 5.05 | 25.50 | 2.25 | 7.11 | 5.55 | 30.80 | 2.36 | 7.45 |
| 5.06 | 25.60 | 2.25 | 7.11 | 5.56 | 30.91 | 2.36 | 7.46 |
| 5.07 | 25.70 | 2.25 | 7.12 | 5.57 | 31.02 | 2.36 | 7.46 |
| 5.08 | 25.81 | 2.25 | 7.13 | 5.58 | 31.14 | 2.36 | 7.47 |
| 5.09 | 25.91 | 2.26 | 7.13 | 5.59 | 31.25 | 2.36 | 7.48 |
| 5.10 | 26.01 | 2.26 | 7.14 | 5.60 | 31.36 | 2.37 | 7.48 |
| 5.11 | 26.11 | 2.26 | 7.15 | 5.61 | 31.47 | 2.37 | 7.49 |
| 5.12 | 26.21 | 2.26 | 7.16 | 5.62 | 31.58 | 2.37 | 7.50 |
| 5.13 | 26.32 | 2.26 | 7.16 | 5.63 | 31.70 | 2.37 | 7.50 |
| 5.14 | 26.42 | 2.27 | 7.17 | 5.64 | 31.81 | 2.37 | 7.51 |
| 5.15 | 26.52 | 2.27 | 7.18 | 5.65 | 31.92 | 2.38 | 7.52 |
| 5.16 | 26.63 | 2.27 | 7.18 | 5.66 | 32.04 | 2.38 | 7.52 |
| 5.17 | 26.73 | 2.27 | 7.19 | 5.67 | 32.15 | 2.38 | 7.53 |
| 5.18 | 26.83 | 2.28 | 7.20 | 5.68 | 32.26 | 2.38 | 7.54 |
| 5.19 | 26.94 | 2.28 | 7.20 | 5.69 | 32.38 | 2.39 | 7.54 |
| 5.20 | 27.04 | 2.28 | 7.21 | 5.70 | 32.49 | 2.39 | 7.55 |
| 5.21 | 27.14 | 2.28 | 7.22 | 5.71 | 32.60 | 2.39 | 7.56 |
| 5.22 | 27.25 | 2.28 | 7.22 | 5.72 | 32.72 | 2.39 | 7.56 |
| 5.23 | 27.35 | 2.29 | 7.23 | 5.73 | 32.83 | 2.39 | 7.57 |
| 5.24 | 27.46 | 2.29 | 7.24 | 5.74 | 32.95 | 2.40 | 7.58 |
| 5.25 | 27.56 | 2.29 | 7.25 | 5.75 | 33.06 | 2.40 | 7.58 |
| 5.26 | 27.67 | 2.29 | 7.25 | 5.76 | 33.18 | 2.40 | 7.59 |
| 5.27 | 27.77 | 2.30 | 7.26 | 5.77 | 33.29 | 2.40 | 7.60 |
| 5.28 | 27.88 | 2.30 | 7.27 | 5.78 | 33.41 | 2.40 | 7.60 |
| 5.29 | 27.98 | 2.30 | 7.27 | 5.79 | 33.52 | 2.41 | 7.61 |
| 5.30 | 28.09 | 2.30 | 7.28 | 5.80 | 33.64 | 2.41 | 7.62 |
| 5.31 | 28.20 | 2.30 | 7.29 | 5.81 | 33.76 | 2.41 | 7.62 |
| 5.32 | 28.30 | 2.31 | 7.29 | 5.82 | 33.87 | 2.41 | 7.63 |
| 5.33 | 28.41 | 2.31 | 7.30 | 5.83 | 33.99 | 2.41 | 7.64 |
| 5.34 | 28.52 | 2.31 | 7.31 | 5.84 | 34.11 | 2.42 | 7.64 |
| 5.35 | 28.62 | 2.31 | 7.31 | 5.85 | 34.22 | 2.42 | 7.65 |
| 5.36 | 28.73 | 2.32 | 7.32 | 5.86 | 34.34 | 2.42 | 7.66 |
| 5.37 | 28.84 | 2.32 | 7.33 | 5.87 | 34.46 | 2.42 | 7.66 |
| 5.38 | 28.94 | 2.32 | 7.33 | 5.88 | 34.57 | 2.42 | 7.67 |
| 5.39 | 29.05 | 2.32 | 7.34 | 5.89 | 34.69 | 2.43 | 7.67 |

**TABLE VI**  TABLE OF SQUARES AND SQUARE ROOTS. (Continued)

| N | $N^2$ | $\sqrt{N}$ | $\sqrt{10N}$ | N | $N^2$ | $\sqrt{N}$ | $\sqrt{10N}$ |
|---|---|---|---|---|---|---|---|
| 5.90 | 34.81 | 2.43 | 7.68 | 6.40 | 40.96 | 2.53 | 8.00 |
| 5.91 | 34.93 | 2.43 | 7.69 | 6.41 | 41.09 | 2.53 | 8.01 |
| 5.92 | 35.05 | 2.43 | 7.69 | 6.42 | 41.22 | 2.53 | 8.01 |
| 5.93 | 35.16 | 2.44 | 7.70 | 6.43 | 41.34 | 2.54 | 8.02 |
| 5.94 | 35.28 | 2.44 | 7.71 | 6.44 | 41.47 | 2.54 | 8.02 |
| 5.95 | 35.40 | 2.44 | 7.71 | 6.45 | 41.60 | 2.54 | 8.03 |
| 5.96 | 35.52 | 2.44 | 7.72 | 6.46 | 41.73 | 2.54 | 8.04 |
| 5.97 | 35.64 | 2.44 | 7.73 | 6.47 | 41.86 | 2.54 | 8.04 |
| 5.98 | 35.76 | 2.45 | 7.73 | 6.48 | 41.99 | 2.55 | 8.05 |
| 5.99 | 35.88 | 2.45 | 7.74 | 6.49 | 42.12 | 2.55 | 8.06 |
| 6.00 | 36.00 | 2.45 | 7.75 | 6.50 | 42.25 | 2.55 | 8.06 |
| 6.01 | 36.12 | 2.45 | 7.75 | 6.51 | 42.38 | 2.55 | 8.07 |
| 6.02 | 36.24 | 2.45 | 7.76 | 6.52 | 42.51 | 2.55 | 8.07 |
| 6.03 | 36.36 | 2.46 | 7.77 | 6.53 | 42.64 | 2.56 | 8.08 |
| 6.04 | 36.48 | 2.46 | 7.77 | 6.54 | 42.77 | 2.56 | 8.09 |
| 6.05 | 36.60 | 2.46 | 7.78 | 6.55 | 42.90 | 2.56 | 8.09 |
| 6.06 | 36.72 | 2.46 | 7.78 | 6.56 | 43.03 | 2.56 | 8.10 |
| 6.07 | 36.84 | 2.46 | 7.79 | 6.57 | 43.16 | 2.56 | 8.11 |
| 6.08 | 36.97 | 2.47 | 7.80 | 6.58 | 43.30 | 2.57 | 8.11 |
| 6.09 | 37.09 | 2.47 | 7.80 | 6.59 | 43.43 | 2.57 | 8.12 |
| 6.10 | 37.21 | 2.47 | 7.81 | 6.60 | 43.56 | 2.57 | 8.12 |
| 6.11 | 37.33 | 2.47 | 7.82 | 6.61 | 43.69 | 2.57 | 8.13 |
| 6.12 | 37.45 | 2.47 | 7.82 | 6.62 | 43.82 | 2.57 | 8.14 |
| 6.13 | 37.58 | 2.48 | 7.83 | 6.63 | 43.96 | 2.57 | 8.14 |
| 6.14 | 37.70 | 2.48 | 7.84 | 6.64 | 44.09 | 2.58 | 8.15 |
| 6.15 | 37.82 | 2.48 | 7.84 | 6.65 | 44.22 | 2.58 | 8.15 |
| 6.16 | 37.95 | 2.48 | 7.85 | 6.66 | 44.36 | 2.58 | 8.16 |
| 6.17 | 38.07 | 2.48 | 7.85 | 6.67 | 44.49 | 2.58 | 8.17 |
| 6.18 | 38.19 | 2.49 | 7.86 | 6.68 | 44.62 | 2.58 | 8.17 |
| 6.19 | 38.32 | 2.49 | 7.87 | 6.69 | 44.76 | 2.59 | 8.18 |
| 6.20 | 38.44 | 2.49 | 7.87 | 6.70 | 44.89 | 2.59 | 8.19 |
| 6.21 | 38.56 | 2.49 | 7.88 | 6.71 | 45.02 | 2.59 | 8.19 |
| 6.22 | 38.69 | 2.49 | 7.89 | 6.72 | 45.16 | 2.59 | 8.20 |
| 6.23 | 38.81 | 2.50 | 7.89 | 6.73 | 45.29 | 2.59 | 8.20 |
| 6.24 | 38.94 | 2.50 | 7.90 | 6.74 | 45.43 | 2.60 | 8.21 |
| 6.25 | 39.06 | 2.50 | 7.91 | 6.75 | 45.56 | 2.60 | 8.22 |
| 6.26 | 39.19 | 2.50 | 7.91 | 6.76 | 45.70 | 2.60 | 8.22 |
| 6.27 | 39.31 | 2.50 | 7.92 | 6.77 | 45.83 | 2.60 | 8.23 |
| 6.28 | 39.44 | 2.51 | 7.92 | 6.78 | 45.97 | 2.60 | 8.23 |
| 6.29 | 39.56 | 2.51 | 7.93 | 6.79 | 46.10 | 2.61 | 8.24 |
| 6.30 | 39.69 | 2.51 | 7.94 | 6.80 | 46.24 | 2.61 | 8.25 |
| 6.31 | 39.82 | 2.51 | 7.94 | 6.81 | 46.38 | 2.61 | 8.25 |
| 6.32 | 39.94 | 2.51 | 7.95 | 6.82 | 46.51 | 2.61 | 8.26 |
| 6.33 | 40.07 | 2.52 | 7.96 | 6.83 | 46.65 | 2.61 | 8.26 |
| 6.34 | 40.20 | 2.52 | 7.96 | 6.84 | 46.79 | 2.62 | 8.27 |
| 6.35 | 40.32 | 2.52 | 7.97 | 6.85 | 46.92 | 2.62 | 8.28 |
| 6.36 | 40.45 | 2.52 | 7.97 | 6.86 | 47.06 | 2.62 | 8.28 |
| 6.37 | 40.58 | 2.52 | 7.98 | 6.87 | 47.20 | 2.62 | 8.29 |
| 6.38 | 40.70 | 2.53 | 7.99 | 6.88 | 47.33 | 2.62 | 8.29 |
| 6.39 | 40.83 | 2.53 | 7.99 | 6.89 | 47.47 | 2.62 | 8.30 |

## APPENDICES

**TABLE VI**  TABLE OF SQUARES AND SQUARE ROOTS. (Continued)

| N | N² | √N | √10N | N | N² | √N | √10N |
|---|---|---|---|---|---|---|---|
| 6.90 | 47.61 | 2.63 | 8.31 | 7.40 | 54.76 | 2.72 | 8.60 |
| 6.91 | 47.75 | 2.63 | 8.31 | 7.41 | 54.91 | 2.72 | 8.61 |
| 6.92 | 47.89 | 2.63 | 8.32 | 7.42 | 55.06 | 2.72 | 8.61 |
| 6.93 | 48.02 | 2.63 | 8.32 | 7.43 | 55.20 | 2.73 | 8.62 |
| 6.94 | 48.16 | 2.63 | 8.33 | 7.44 | 55.35 | 2.73 | 8.63 |
| 6.95 | 48.30 | 2.64 | 8.34 | 7.45 | 55.50 | 2.73 | 8.63 |
| 6.96 | 48.44 | 2.64 | 8.34 | 7.46 | 55.65 | 2.73 | 8.64 |
| 6.97 | 48.58 | 2.64 | 8.35 | 7.47 | 55.80 | 2.73 | 8.64 |
| 6.98 | 48.72 | 2.64 | 8.35 | 7.48 | 55.95 | 2.73 | 8.65 |
| 6.99 | 48.86 | 2.64 | 8.36 | 7.49 | 56.10 | 2.74 | 8.65 |
| 7.00 | 49.00 | 2.65 | 8.37 | 7.50 | 56.25 | 2.74 | 8.66 |
| 7.01 | 49.14 | 2.65 | 8.37 | 7.51 | 56.40 | 2.74 | 8.67 |
| 7.02 | 49.28 | 2.65 | 8.38 | 7.52 | 56.55 | 2.74 | 8.67 |
| 7.03 | 49.42 | 2.65 | 8.38 | 7.53 | 56.70 | 2.74 | 8.68 |
| 7.04 | 49.56 | 2.65 | 8.39 | 7.54 | 56.85 | 2.75 | 8.68 |
| 7.05 | 49.70 | 2.66 | 8.40 | 7.55 | 57.00 | 2.75 | 8.69 |
| 7.06 | 49.84 | 2.66 | 8.40 | 7.56 | 57.15 | 2.75 | 8.69 |
| 7.07 | 49.98 | 2.66 | 8.41 | 7.57 | 57.30 | 2.75 | 8.70 |
| 7.08 | 50.13 | 2.66 | 8.41 | 7.58 | 57.46 | 2.75 | 8.71 |
| 7.09 | 50.27 | 2.66 | 8.42 | 7.59 | 57.61 | 2.75 | 8.71 |
| 7.10 | 50.41 | 2.66 | 8.43 | 7.60 | 57.76 | 2.76 | 8.72 |
| 7.11 | 50.55 | 2.67 | 8.43 | 7.61 | 57.91 | 2.76 | 8.72 |
| 7.12 | 50.69 | 2.67 | 8.44 | 7.62 | 58.06 | 2.76 | 8.73 |
| 7.13 | 50.84 | 2.67 | 8.44 | 7.63 | 58.22 | 2.76 | 8.73 |
| 7.14 | 50.98 | 2.67 | 8.45 | 7.64 | 58.37 | 2.76 | 8.74 |
| 7.15 | 51.12 | 2.67 | 8.46 | 7.65 | 58.52 | 2.77 | 8.75 |
| 7.16 | 51.27 | 2.68 | 8.46 | 7.66 | 58.68 | 2.77 | 8.75 |
| 7.17 | 51.41 | 2.68 | 8.47 | 7.67 | 58.83 | 2.77 | 8.76 |
| 7.18 | 51.55 | 2.68 | 8.47 | 7.68 | 58.98 | 2.77 | 8.76 |
| 7.19 | 51.70 | 2.68 | 8.48 | 7.69 | 59.14 | 2.77 | 8.77 |
| 7.20 | 51.84 | 2.68 | 8.49 | 7.70 | 59.29 | 2.77 | 8.77 |
| 7.21 | 51.98 | 2.69 | 8.49 | 7.71 | 59.44 | 2.78 | 8.78 |
| 7.22 | 52.13 | 2.69 | 8.50 | 7.72 | 59.60 | 2.78 | 8.79 |
| 7.23 | 52.27 | 2.69 | 8.50 | 7.73 | 59.75 | 2.78 | 8.79 |
| 7.24 | 52.42 | 2.69 | 8.51 | 7.74 | 59.91 | 2.78 | 8.80 |
| 7.25 | 52.56 | 2.69 | 8.51 | 7.75 | 60.06 | 2.78 | 8.80 |
| 7.26 | 52.71 | 2.69 | 8.52 | 7.76 | 60.22 | 2.79 | 8.81 |
| 7.27 | 52.85 | 2.70 | 8.53 | 7.77 | 60.37 | 2.79 | 8.81 |
| 7.28 | 53.00 | 2.70 | 8.53 | 7.78 | 60.53 | 2.79 | 8.82 |
| 7.29 | 53.14 | 2.70 | 8.54 | 7.79 | 60.68 | 2.79 | 8.83 |
| 7.30 | 53.29 | 2.70 | 8.54 | 7.80 | 60.84 | 2.79 | 8.83 |
| 7.31 | 53.44 | 2.70 | 8.55 | 7.81 | 61.00 | 2.79 | 8.84 |
| 7.32 | 53.58 | 2.71 | 8.56 | 7.82 | 61.15 | 2.80 | 8.84 |
| 7.33 | 53.73 | 2.71 | 8.56 | 7.83 | 61.31 | 2.80 | 8.85 |
| 7.34 | 53.88 | 2.71 | 8.57 | 7.84 | 61.47 | 2.80 | 8.85 |
| 7.35 | 54.02 | 2.71 | 8.57 | 7.85 | 61.62 | 2.80 | 8.86 |
| 7.36 | 54.17 | 2.71 | 8.58 | 7.86 | 61.78 | 2.80 | 8.87 |
| 7.37 | 54.32 | 2.71 | 8.58 | 7.87 | 61.94 | 2.81 | 8.87 |
| 7.38 | 54.46 | 2.72 | 8.59 | 7.88 | 62.09 | 2.81 | 8.88 |
| 7.39 | 54.61 | 2.72 | 8.60 | 7.89 | 62.25 | 2.81 | 8.88 |

**TABLE VI** TABLE OF SQUARES AND SQUARE ROOTS. (Continued)

| N | N² | √N | √10N | N | N² | √N | √10N |
|---|---|---|---|---|---|---|---|
| 7.90 | 62.41 | 2.81 | 8.89 | 8.40 | 70.56 | 2.90 | 9.17 |
| 7.91 | 62.57 | 2.81 | 8.89 | 8.41 | 70.73 | 2.90 | 9.17 |
| 7.92 | 62.73 | 2.81 | 8.90 | 8.42 | 70.90 | 2.90 | 9.18 |
| 7.93 | 62.88 | 2.82 | 8.91 | 8.43 | 71.06 | 2.90 | 9.18 |
| 7.94 | 63.04 | 2.82 | 8.91 | 8.44 | 71.23 | 2.91 | 9.19 |
| 7.95 | 63.20 | 2.82 | 8.92 | 8.45 | 71.40 | 2.91 | 9.19 |
| 7.96 | 63.36 | 2.82 | 8.92 | 8.46 | 71.57 | 2.91 | 9.20 |
| 7.97 | 63.52 | 2.82 | 8.93 | 8.47 | 71.74 | 2.91 | 9.20 |
| 7.98 | 63.68 | 2.82 | 8.93 | 8.48 | 71.91 | 2.91 | 9.21 |
| 7.99 | 63.84 | 2.83 | 8.94 | 8.49 | 72.08 | 2.91 | 9.21 |
| 8.00 | 64.00 | 2.83 | 8.94 | 8.50 | 72.25 | 2.92 | 9.22 |
| 8.01 | 64.16 | 2.83 | 8.95 | 8.51 | 72.42 | 2.92 | 9.22 |
| 8.02 | 64.32 | 2.83 | 8.96 | 8.52 | 72.59 | 2.92 | 9.23 |
| 8.03 | 64.48 | 2.83 | 8.96 | 8.53 | 72.76 | 2.92 | 9.24 |
| 8.04 | 64.64 | 2.84 | 8.97 | 8.54 | 72.93 | 2.92 | 9.24 |
| 8.05 | 64.80 | 2.84 | 8.97 | 8.55 | 73.10 | 2.92 | 9.25 |
| 8.06 | 64.96 | 2.84 | 8.98 | 8.56 | 73.27 | 2.93 | 9.25 |
| 8.07 | 65.12 | 2.84 | 8.98 | 8.57 | 73.44 | 2.93 | 9.26 |
| 8.08 | 65.29 | 2.84 | 8.99 | 8.58 | 73.62 | 2.93 | 9.26 |
| 8.09 | 65.45 | 2.84 | 8.99 | 8.59 | 73.79 | 2.93 | 9.27 |
| 8.10 | 65.61 | 2.85 | 9.00 | 8.60 | 73.96 | 2.93 | 9.27 |
| 8.11 | 65.77 | 2.85 | 9.01 | 8.61 | 74.13 | 2.93 | 9.28 |
| 8.12 | 65.93 | 2.85 | 9.01 | 8.62 | 74.30 | 2.94 | 9.28 |
| 8.13 | 66.10 | 2.85 | 9.02 | 8.63 | 74.48 | 2.94 | 9.29 |
| 8.14 | 66.26 | 2.85 | 9.02 | 8.64 | 74.65 | 2.94 | 9.30 |
| 8.15 | 66.42 | 2.85 | 9.03 | 8.65 | 74.82 | 2.94 | 9.30 |
| 8.16 | 66.59 | 2.86 | 9.03 | 8.66 | 75.00 | 2.94 | 9.31 |
| 8.17 | 66.75 | 2.86 | 9.04 | 8.67 | 75.17 | 2.94 | 9.31 |
| 8.18 | 66.91 | 2.86 | 9.04 | 8.68 | 75.34 | 2.95 | 9.32 |
| 8.19 | 67.08 | 2.86 | 9.05 | 8.69 | 75.52 | 2.95 | 9.32 |
| 8.20 | 67.24 | 2.86 | 9.06 | 8.70 | 75.69 | 2.95 | 9.33 |
| 8.21 | 67.40 | 2.87 | 9.06 | 8.71 | 75.86 | 2.95 | 9.33 |
| 8.22 | 67.57 | 2.87 | 9.07 | 8.72 | 76.04 | 2.95 | 9.34 |
| 8.23 | 67.73 | 2.87 | 9.07 | 8.73 | 76.21 | 2.95 | 9.34 |
| 8.24 | 67.90 | 2.87 | 9.08 | 8.74 | 76.39 | 2.96 | 9.35 |
| 8.25 | 68.06 | 2.87 | 9.08 | 8.75 | 76.56 | 2.96 | 9.35 |
| 8.26 | 68.23 | 2.87 | 9.09 | 8.76 | 76.74 | 2.96 | 9.36 |
| 8.27 | 68.39 | 2.88 | 9.09 | 8.77 | 76.91 | 2.96 | 9.36 |
| 8.28 | 68.56 | 2.88 | 9.10 | 8.78 | 77.09 | 2.96 | 9.37 |
| 8.29 | 68.72 | 2.88 | 9.10 | 8.79 | 77.26 | 2.96 | 9.38 |
| 8.30 | 68.89 | 2.88 | 9.11 | 8.80 | 77.44 | 2.97 | 9.38 |
| 8.31 | 69.06 | 2.88 | 9.12 | 8.81 | 77.62 | 2.97 | 9.39 |
| 8.32 | 69.22 | 2.88 | 9.12 | 8.82 | 77.79 | 2.97 | 9.39 |
| 8.33 | 69.39 | 2.89 | 9.13 | 8.83 | 77.97 | 2.97 | 9.40 |
| 8.34 | 69.56 | 2.89 | 9.13 | 8.84 | 78.15 | 2.97 | 9.40 |
| 8.35 | 69.72 | 2.89 | 9.14 | 8.85 | 78.32 | 2.97 | 9.41 |
| 8.36 | 69.89 | 2.89 | 9.14 | 8.86 | 78.50 | 2.98 | 9.41 |
| 8.37 | 70.06 | 2.89 | 9.15 | 8.87 | 78.68 | 2.98 | 9.42 |
| 8.38 | 70.22 | 2.89 | 9.15 | 8.88 | 78.85 | 2.98 | 9.42 |
| 8.39 | 70.39 | 2.90 | 9.16 | 8.89 | 79.03 | 2.98 | 9.43 |

## APPENDICES

**TABLE VI** TABLE OF SQUARES AND SQUARE ROOTS. (Continued)

| N | $N^2$ | $\sqrt{N}$ | $\sqrt{10N}$ | N | $N^2$ | $\sqrt{N}$ | $\sqrt{10N}$ |
|---|---|---|---|---|---|---|---|
| 8.90 | 79.21 | 2.98 | 9.43 | 9.40 | 88.36 | 3.07 | 9.70 |
| 8.91 | 79.39 | 2.98 | 9.44 | 9.41 | 88.55 | 3.07 | 9.70 |
| 8.92 | 79.57 | 2.99 | 9.44 | 9.42 | 88.74 | 3.07 | 9.71 |
| 8.93 | 79.74 | 2.99 | 9.45 | 9.43 | 88.92 | 3.07 | 9.71 |
| 8.94 | 79.92 | 2.99 | 9.46 | 9.44 | 89.11 | 3.07 | 9.72 |
| 8.95 | 80.10 | 2.99 | 9.46 | 9.45 | 89.30 | 3.07 | 9.72 |
| 8.96 | 80.28 | 2.99 | 9.47 | 9.46 | 89.49 | 3.08 | 9.73 |
| 8.97 | 80.46 | 2.99 | 9.47 | 9.47 | 89.68 | 3.08 | 9.73 |
| 8.98 | 80.64 | 3.00 | 9.48 | 9.48 | 89.87 | 3.08 | 9.74 |
| 8.99 | 80.82 | 3.00 | 9.48 | 9.49 | 90.06 | 3.08 | 9.74 |
| 9.00 | 81.00 | 3.00 | 9.49 | 9.50 | 90.25 | 3.08 | 9.75 |
| 9.01 | 81.18 | 3.00 | 9.49 | 9.51 | 90.44 | 3.08 | 9.75 |
| 9.02 | 81.36 | 3.00 | 9.50 | 9.52 | 90.63 | 3.09 | 9.76 |
| 9.03 | 81.54 | 3.00 | 9.50 | 9.53 | 90.82 | 3.09 | 9.76 |
| 9.04 | 81.72 | 3.01 | 9.51 | 9.54 | 91.01 | 3.09 | 9.77 |
| 9.05 | 81.90 | 3.01 | 9.51 | 9.55 | 91.20 | 3.09 | 9.77 |
| 9.06 | 82.08 | 3.01 | 9.52 | 9.56 | 91.39 | 3.09 | 9.78 |
| 9.07 | 82.26 | 3.01 | 9.52 | 9.57 | 91.58 | 3.09 | 9.78 |
| 9.08 | 82.45 | 3.01 | 9.53 | 9.58 | 91.78 | 3.10 | 9.79 |
| 9.09 | 82.63 | 3.01 | 9.53 | 9.59 | 91.97 | 3.10 | 9.79 |
| 9.10 | 82.81 | 3.02 | 9.54 | 9.60 | 92.16 | 3.10 | 9.80 |
| 9.11 | 82.99 | 3.02 | 9.54 | 9.61 | 92.35 | 3.10 | 9.80 |
| 9.12 | 83.17 | 3.02 | 9.55 | 9.62 | 92.54 | 3.10 | 9.81 |
| 9.13 | 83.36 | 3.02 | 9.56 | 9.63 | 92.74 | 3.10 | 9.81 |
| 9.14 | 83.54 | 3.02 | 9.56 | 9.64 | 92.93 | 3.10 | 9.82 |
| 9.15 | 83.72 | 3.02 | 9.57 | 9.65 | 93.12 | 3.11 | 9.82 |
| 9.16 | 83.91 | 3.03 | 9.57 | 9.66 | 93.32 | 3.11 | 9.83 |
| 9.17 | 84.09 | 3.03 | 9.58 | 9.67 | 93.51 | 3.11 | 9.83 |
| 9.18 | 84.27 | 3.03 | 9.58 | 9.68 | 93.70 | 3.11 | 9.84 |
| 9.19 | 84.46 | 3.03 | 9.59 | 9.69 | 93.90 | 3.11 | 9.84 |
| 9.20 | 84.64 | 3.03 | 9.59 | 9.70 | 94.09 | 3.11 | 9.85 |
| 9.21 | 84.82 | 3.03 | 9.60 | 9.71 | 94.28 | 3.12 | 9.85 |
| 9.22 | 85.01 | 3.04 | 9.60 | 9.72 | 94.48 | 3.12 | 9.86 |
| 9.23 | 85.19 | 3.04 | 9.61 | 9.73 | 94.67 | 3.12 | 9.86 |
| 9.24 | 85.38 | 3.04 | 9.61 | 9.74 | 94.87 | 3.12 | 9.87 |
| 9.25 | 85.56 | 3.04 | 9.62 | 9.75 | 95.06 | 3.12 | 9.87 |
| 9.26 | 85.75 | 3.04 | 9.62 | 9.76 | 95.26 | 3.12 | 9.88 |
| 9.27 | 85.93 | 3.04 | 9.63 | 9.77 | 95.45 | 3.13 | 9.88 |
| 9.28 | 86.12 | 3.05 | 9.63 | 9.78 | 95.65 | 3.13 | 9.89 |
| 9.29 | 86.30 | 3.05 | 9.64 | 9.79 | 95.84 | 3.13 | 9.89 |
| 9.30 | 86.49 | 3.05 | 9.64 | 9.80 | 96.04 | 3.13 | 9.90 |
| 9.31 | 86.68 | 3.05 | 9.65 | 9.81 | 96.24 | 3.13 | 9.90 |
| 9.32 | 86.86 | 3.05 | 9.65 | 9.82 | 96.43 | 3.13 | 9.91 |
| 9.33 | 87.05 | 3.05 | 9.66 | 9.83 | 96.63 | 3.14 | 9.91 |
| 9.34 | 87.24 | 3.06 | 9.66 | 9.84 | 96.83 | 3.14 | 9.92 |
| 9.35 | 87.42 | 3.06 | 9.67 | 9.85 | 97.02 | 3.14 | 9.92 |
| 9.36 | 87.61 | 3.06 | 9.67 | 9.86 | 97.22 | 3.14 | 9.93 |
| 9.37 | 87.80 | 3.06 | 9.68 | 9.87 | 97.42 | 3.14 | 9.93 |
| 9.38 | 87.98 | 3.06 | 9.69 | 9.88 | 97.61 | 3.14 | 9.94 |
| 9.39 | 88.17 | 3.06 | 9.69 | 9.89 | 97.81 | 3.14 | 9.94 |

**TABLE VI** TABLE OF SQUARES AND SQUARE ROOTS. (Continued)

| $N$ | $N^2$ | $\sqrt{N}$ | $\sqrt{10N}$ | $N$ | $N^2$ | $\sqrt{N}$ | $\sqrt{10N}$ |
|---|---|---|---|---|---|---|---|
| 9.90 | 98.01 | 3.15 | 9.95 | 9.95 | 99.00 | 3.15 | 9.97 |
| 9.91 | 98.21 | 3.15 | 9.95 | 9.96 | 99.20 | 3.16 | 9.98 |
| 9.92 | 98.41 | 3.15 | 9.96 | 9.97 | 99.40 | 3.16 | 9.98 |
| 9.93 | 98.60 | 3.15 | 9.96 | 9.98 | 99.60 | 3.16 | 9.99 |
| 9.94 | 98.80 | 3.15 | 9.97 | 9.99 | 99.80 | 3.16 | 9.99 |
|  |  |  |  | 10.00 | 100.00 | 3.16 | 10.00 |

EXAMPLES.

$\sqrt{1.23} = 1.11$ ...... Find value under radical in first column; read root in third column

$\sqrt{12.3} = 3.51$ ...... Find numerical portion of value under radical in first column; read root in fourth column

$\sqrt{123} = 11.1$
$\sqrt{1230} = 35.1$
$\sqrt{0.123} = .351$
$\sqrt{0.0123} = 0.111$
$12.3^2 = 151$
$1.23^2 = 1.51$

......... If the number under the radical is not a tabled value of $N$ or $10N$, move the decimal in groups of two places until it is a tabled value of $N$ or $10N$. Read root in columns three or four and then move decimal in result one place for each group of two places moved in original number.

## APPENDICES

**TABLE VII** TABLE OF CRITICAL VALUES FOR USE WITH THE WALD-WOLFOWITZ RUNS TEST (ADAPTED FROM TABLE II). A NUMBER OF RUNS AS SMALL AS OR SMALLER THAN THE VALUE GIVEN IN THE TABLE WOULD OCCUR BY CHANCE LESS THAN 5% OF THE TIME IF THERE WERE NO TRUE DIFFERENCE BETWEEN THE TWO DISTRIBUTIONS SAMPLED.

| $n_2$ \ $n_1$ | 2 | 3 | 4 | 5 | 6 | 7 | 8 | 9 | 10 |
|---|---|---|---|---|---|---|---|---|---|
| 2 | | | | | | | 2 | 2 | 2 |
| 3 | | | | 2 | 2 | 2 | 2 | 2 | 3 |
| 4 | | | 2 | 2 | 3 | 3 | 3 | 3 | 3 |
| 5 | | 2 | 2 | 3 | 3 | 3 | 3 | 4 | 4 |
| 6 | | 2 | 3 | 3 | 3 | 4 | 4 | 4 | 5 |
| 7 | | 2 | 3 | 3 | 4 | 4 | 4 | 5 | 5 |
| 8 | 2 | 2 | 3 | 3 | 4 | 4 | 5 | 5 | 6 |
| 9 | 2 | 2 | 3 | 4 | 4 | 5 | 5 | 6 | 6 |
| 10 | 2 | 3 | 3 | 4 | 5 | 5 | 6 | 6 | 6 |

EXAMPLE. To test whether the following two groups of observations differ in their measures of central tendency, dispersion, skewness, or kurtosis:

| A | 1 | 2 | 2 | 3 | 4 | 4 | 4 | 6 | 8 | 10 |
|---|---|---|---|---|---|---|---|---|---|---|
| B | 5 | 5 | 9 | 13 | 14 | 19 | 22 | 25 | 30 | 35 |

1. *Ordered data:* <u>1 2 2 3 4 4 4</u> <u>5 5</u> <u>6 8 9 10</u> <u>13 14 19 22 25 30 35</u>
2. *#Runs* = 6, $n_1 = 10$, $n_2 = 10$.
3. *Tabled value* = 6. Since the sample number of runs is the same as the tabled value, the conclusion is that, based on this sample data, there are differences between the two samples. In other words, less than 5% of the time would there be this few runs because of sampling error if in fact there were no true differences between the two distributions.

**TABLE VIII** CRITICAL VALUES OF THE CHI-SQUARE DISTRIBUTION. VALUES, CALCULATED FROM SAMPLE DATA, THAT ARE AS LARGE AS OR LARGER THAN THOSE LISTED IN THE TABLE OCCUR WITH THE PROBABILITY GIVEN BY THE COLUMN HEADINGS.

| df | .25 | .10 | .05 | .01 | .005 |
|---|---|---|---|---|---|
| 1 | 1.32 | 2.71 | 3.84 | 6.63 | 7.88 |
| 2 | 2.77 | 4.61 | 5.99 | 9.21 | 10.6 |
| 3 | 4.11 | 6.25 | 7.81 | 11.3 | 12.8 |
| 4 | 5.39 | 7.78 | 9.49 | 13.3 | 14.9 |
| 5 | 6.63 | 9.24 | 11.1 | 15.1 | 16.7 |
| 6 | 7.84 | 10.6 | 12.6 | 16.8 | 18.5 |
| 7 | 9.04 | 12.0 | 14.1 | 18.5 | 20.3 |
| 8 | 10.2 | 13.4 | 15.5 | 20.1 | 22.0 |
| 9 | 11.4 | 14.7 | 16.9 | 21.7 | 23.6 |
| 10 | 12.5 | 16.0 | 18.3 | 23.2 | 25.2 |
| 11 | 13.7 | 17.3 | 19.7 | 24.7 | 26.8 |
| 12 | 14.8 | 18.5 | 21.0 | 26.2 | 28.3 |
| 13 | 16.0 | 19.8 | 22.4 | 27.7 | 29.8 |
| 14 | 17.1 | 21.1 | 23.7 | 29.1 | 31.3 |
| 15 | 18.2 | 22.3 | 25.0 | 30.6 | 32.8 |

EXAMPLE. Suppose a die were tossed 72 times and 10 ones were observed, along with 15 twos, 11 threes, 12 fours, 8 fives, and 16 sixes. Is the die fair? If the die is fair, then one would expect 12 of each number to occur. Thus we have

| Number | 1 | 2 | 3 | 4 | 5 | 6 |
|---|---|---|---|---|---|---|
| Observed | 10 | 15 | 11 | 12 | 8 | 16 |
| Expected | 12 | 12 | 12 | 12 | 12 | 12 |

$$\chi_5^2 = \frac{(10-12)^2}{12} + \frac{(15-12)^2}{12} + \frac{(11-12)^2}{12} + \frac{(12-12)^2}{12} + \frac{(8-12)^2}{12}$$

$$+ \frac{(16-12)^2}{12} = \frac{46}{12} = 3.833, \text{ which by the table is not extreme.}$$

Conclusion: Based on the sample data, we would conclude that the die is fair.

# INDEX

absolute value, 140, 142, 147
accidental occurrences, 174
attribute, 201
average, 19–30, 175
   see also mean
average deviation, 138, 142, 147

bar chart, 6, 7
before-after design, 92, 93, 100, 102, 201
Bernoulli distribution, 65–71
Bernoulli trial, 66, 70, 73, 77
bias, 31, 69, 70, 71
biased sample, 36, 44
binomial
   distribution, 72–84, 180, 224
   expansion, 73, 76, 85
   experiment, 76, 86
   formula, 77
   model, 87–89
   probabilities, 248–249
   test, 85–91
blind experiment, 70, 71

category, 4, 13, 14, 50, 120, 186, 202
center of gravity, 163
Central Limit Theorem, 210, 227
central sample moments, 165–166
central tendency, 19, 30, 161, 171
   see also measures of central tendency
chance, 33
   fluctuation, 101, 170

chance (continued)
   of error, 52
chart, bar, 6, 7
   pie, 3, 9–10
chi-square
   distribution, 184–189, 196–198, 263
   statistic, 200
   test, 184, 196–197, 200, 223
classification, 13, 96, 189, 196
clumping, 51–54, 172
coefficient, 76, 78
combinations, 112–115, 120–122
compound event, 55, 58, 64
counting, 120
counting rule (formula), 108–112, 114, 119–122, 125, 132
critical point, 86–88, 101, 102, 106, 132, 187, 198, 223
critical value, see critical point
cutoff point (value), see critical point

decision, 2, 3, 87, 220–223
defective, 51, 85, 88, 89, 180
degree of precision, see precision
degrees of freedom, 185, 189, 196, 200
degrees of skewness, see skewness
dependent experiments, 123, 126, 129–130
dependent pairs, see matched subjects
descriptive statistics, 3, 4, 12, 205
design, before-and-after, 92, 93, 100, 102, 201
deviations, 142–143, 147, 165, 202
disjoint, 66, 70

# INDEX

dispersion, 137, 150, 161, 170–172
distribution, 13–18, 56, 60–61, 63, 64
   see also under name of distribution

empirical distribution, see distribution
equally likely, 125
error, 52, 138, 223, 227
   see also sampling error
estimate, 39, 205–206
estimation, 205–207, 209–213, 218–219
event, 32–36, 40, 44, 66, 77, 171
   compound, 55, 58, 64
   simple, 55, 58, 64, 124
   see also outcome
expansion, binomial, 73, 76, 85
expected frequency, 56, 189, 198, 202
expected results, 184
experiment, 2, 3, 12, 31–33, 47, 55, 63, 66, 72, 93, 123, 207
   binomial, 76, 86
   blind, 70, 71
   dependent, 123, 126, 129–130
   independent, 123–126, 129–130
experimental bias, 69
experimental treatment, 47
   see also treatment
extreme (event, observation, outcome) 86–88, 91, 95, 101–104

factor, 92, 101, 102
factorial form, 108, 111, 114, 119
factorial notation, 107, 108, 119
failure, 72, 73, 84
formula, binomial, 77
   counting, see counting rule
freedom, degrees of, 185, 189, 196, 200
frequency, 6, 12, 13
   distribution, 157–158
   expected, 56, 189, 198, 202
   observed, 189, 198
   table, 13, 14, 18

graph, 6, 158
gravity, center of, 163
grouped data, 20, 164

histogram, 6–8, 20, 176
hypothesis, 221

impossible event, 57, 58
independence, 72, 84, 186, 196, 198, 201
independent events, see independence
independent experiments, 123–126, 129–130
independent groups, 131
independent measures, 130
independent outcomes, 123

independent samples, 131
inductive statistics, 3
inference, 3, 4, 34, 205
interval data (measurement), see interval scale
interval scale, 4, 6, 23, 102, 142

kurtosis, 159, 161, 164, 169–172

Law of Large Numbers, 39–44, 206
left-skewed, see skewness
leptokurtic, 159, 164, 169

Mann-Whitney $U$ Test, 131, 134–135, 151, 201, 202, 223, 251
marginal total, 196, 198
matched subjects (individuals, pairs), 93, 100, 102, 201
mathematical model, 56
   see also model
mean, 19, 22–24, 28, 38–39, 101, 143, 157, 162–163, 170, 205, 207, 224
mean deviation, see average deviation
measure, independent, 130
   of central tendency, 20, 29, 138, 148, 157, 159, 162, 201, 223
      see also averages
   of dispersion, 138, 143, 148, 157, 162, 166
   of error, see error
   of location, see measure of central tendency
measurement, 2, 12, 33, 39, 69–70
measurement scale, 4–5, 12, 93, 102, 131
median, 20–21, 23–24, 28, 30, 138, 142, 163, 165, 188
median test, 201–203, 223
mesokurtic, 159, 164, 169, 224
midrange, 22–24, 30, 138
mode, 21–24, 27, 29, 30, 101, 163
model, assignment of, 40, 187, 189
   binomial (coin-toss), 65, 74, 81, 83, 86–89
   definition of, 36–37, 44
   die-toss, 61–63
   mathematical, 56–58, 184, 202, 223
   normal distribution, 224
   Poisson, 176
moments, see sample moments
mutually exclusive, 66, 70

negatively skewed, see skewness
nominal scale (data, measure), 4, 6, 9, 15, 22–23, 93, 102, 131, 201
nonrandom, 36, 53–54
normal distribution, 159, 224, 227, 230, 236

observations, 2, 12, 19, 38, 205
observed frequency, 189, 198

# INDEX

observed results, 184
order, 108–113
ordinal scale (data, measure), 4, 23, 93, 102, 142, 201, 202
outcome, 31, 33, 40, 72, 74, 77, 101, 111, 123, 125
   *see also* event
outlier, 19, 22–24, 30, 51, 142
outlying observation, *see* outlier

paired measurements, *see* matched subjects
Pascal Triangle, 73, 75–77, 83, 118–119
peakedness, 159
Pearsonian coefficient of skewness, 162–163, 166
percentage, 10, 102, 104
   *see also* proportion, probability
permutations, 107–114, 119–122
   *see also* proportion, probability
permute, 110–119
   *see also* permutations
pie chart, 3, 9–10
   platykurtic, 159, 169
Poisson distribution, 174–176, 180–183, 206
Poisson events, 174–175
Poisson, S. D., 174
polygon, 6–8
population, 31–34, 38, 43, 45, 51–52, 55, 171, 205, 223
positively skewed, *see* skewness
precision, 138, 205, 210
predict, 27
probability
   binomial, 248–249
   calculation of, 36, 37
      from binomial formula, 77, 85–86
      from Pascal Triangle, 73–74, 77
   definition of, 33, 44
   distribution, *see* distribution
   in chi-square test, 189
   in coin toss, 72–74, 83
   in die toss, 56
   in lottery, 57
   in sign test, 94–95
   in signed ranks test, 103–104
   of error, *see* error
   properties of, 58
production control, 89
properties of probability, 57–58
proportion, 9–10, 34, 37, 39, 205, 207
   *see also* probability, percentage

quality control, 51, 180
questionnaire, 3

random fluctuations, see chance fluctuations
random numbers (digits), 48, 63
   table of, 46, 49, 63, 246–247
random order, 46, 49–50
random sample, 34, 45, 52

random sequence, 46, 49, 52, 223
randomness, 45–49, 51
   test of, 50–54
range, 138, 147
rank, 4, 104, 131, 148–149
   signed, *see* signed ranks test
rare event, 86–87, 91, 180, 184
relationship, 197
relative frequency, 9, 12, 61, 80–81, 83, 207
representative sample, 51–52
response, 92, 130
right-skewed, *see* skewness
run, 50–54, 129, 170–171

sample, 31–32, 39, 43
   biased, 36, 44
   evidence, 94
   independent, 131
   size, 94, 205
sample moments, 161, 169
   central, 165–166
sample points, 124
sample space, 32–33, 35, 44, 57, 59
sampled population, 34–35, 43–44
sampling, 31–44, 51
   with replacement, 32, 37–38, 42, 44
   without replacement, 32, 35–36, 40, 42, 44
sampling distribution, 56, 176
sampling error, 40, 56, 62–64, 170, 187
   *see also* error
sampling urn, 176
sampling variations, 184
sequence, 50, 51, 54
$\Sigma$ (sigma), 139, 147
sign test, 92–95, 102–103, 130, 201, 223
signed ranks test, 101–106, 130–132, 201, 223, 250
significantly better, 92
simple event, 55, 58, 64, 124
simple random sample, 35, 37, 44–46
size of sample, *see* sample size
skewed distribution, *see* skewness
skewness, 158–159, 161, 163, 166, 169, 170–172, 227
square root, 140, 143, 252–261
standard deviation, 138, 143, 147, 157, 162, 166, 206–207, 224
standard error, 210, 219
statistics, 1, 3, 12, 31, 101, 106, 220
statisticians, 2
success, 72–73, 84
survey, 2–3
symmetric distribution, *see* symmetry
symmetry, 157, 159, 169, 224, 227

table of random digits, *see* random numbers table
tally, *see* work mark
target population, 34–35, 43–44
test, binomial, 85–91
   chi-square, 184, 196–197, 200, 223

267

## INDEX

test *(continued)*
    Mann-Whitney $U$, 131, 134–135, 151, 201, 223, 251
    median, 201–203, 223
    of fit, 188
    runs (Wald-Wolfowitz), 170–171, 223, 262
    sign, 92–95, 102–103, 130, 201, 223
    signed ranks, 101–106, 130–132, 201, 223, 250
test statistic, 223
    *see also* statistics
theoretical distribution, 57, 65
tied ranks, 131, 150
treatment, 55, 63, 69, 92, 96, 100–101
treatment effects, 130
treatment groups, 47
tree diagram, 109–110, 123
trial, Bernoulli, 66, 70, 73, 77
tri-modal, 22

uniform distribution, 55–65, 72, 182, 186
unimodal, 163

unpredictable, 53
unusual, 86
variability, 39, 205
variance, 143, 147, 162

Wald-Wolfowitz Runs Test, 170–171, 223, 262
Wilcoxon test, *see* signed ranks test
with replacement, *see* sampling with replacement
without replacement, *see* sampling without replacement
work marks, 62

$\bar{x}$, *see* mean
$\tilde{x}$, *see* median